# MINDS, BRAINS AND SCIENCE

# MINDS, BRAINS AND SCIENCE

JOHN SEARLE

HARVARD UNIVERSITY PRESS
Cambridge, Massachusetts
1984

Printed in the United States of America
10 9 8 7 6 5 4 3 2 1

This book is printed on acid-free paper, and its binding materials have been chosen for strength and durability.

Library of Congress Cataloging in Publication Data

Searle, John R.
Minds, brains and science.

(The 1984 Reith lectures)
Bibliography: p.
Includes index.
1. Mind and body. 2. Brain. 3. Thought and thinking.
I. Title. II. Series: Reith lectures; 1984.
BF161.S352 1985 128'.2 84-25260
ISBN 0-674-57631-4

# CONTENTS

For Dagmar

# INTRODUCTION

It was a great honour for me to be asked to give the 1984 Reith Lectures. Since Bertrand Russell began the series in 1948, these are the first to be given by a philosopher.

But if to give the lectures is an honour, it is also a challenge. The ideal series of Reith lectures should consist of six broadcast units, each of exactly one half hour in length, each a self-contained entity that can stand on its own, yet each contributing to a unified whole consisting of the six. The series should build on the previous work of the lecturer, but at the same time it should contain new and original material. And, perhaps hardest of all to achieve, it should be completely accessible to an interested and alert audience most of whose members have no familiarity whatever with the subject matter, with its terminology, or with the special preoccupations of its practitioners. I do not know if all of these objectives are simultaneously achievable, but at any rate they are what I was aiming at. One of my strongest reasons for wanting to give the Reith lectures was the conviction that the results and methods of modern analytic philosophy can be made available to a much wider audience.

My first plans for the book version were to expand each of the chapters in a way that would attempt to meet all of the objections that I could imagine coming from my cantankerous fellow philosophers, not to mention colleagues in cognitive science, artificial intelligence, and other fields. My original plan, in short, was to try to convert the lectures into a conventional book with footnotes and all the rest of it. In the end I decided against that precisely because it would destroy what to me was one of the most appealing things about the series in the first place: its complete accessibility to anybody who is interested enough to try to follow the arguments. These chapters, then, are essentially the Reith lectures as I delivered them. I have expanded some in the interests of greater clarity, but I have tried to keep the style and tone and informality of the original lectures.

The overriding theme of the series concerns the relationship of human beings to the rest of the universe. Specifically, it concerns the question of how we reconcile a certain traditional mentalistic conception that we have of ourselves with an apparently inconsistent conception of the universe as a purely physical system, or a set of interacting physical systems. Around this theme, each chapter is addressed to a specific question: what is the relation of the mind to the brain? Can digital computers have minds solely in virtue of having the right programs with the right inputs and outputs? How plausible is the model of the mind as a computer program? What is the nature of the structure of human action? What is the status of the social sciences as sciences? How, if at all, can we reconcile our conviction of our own free will with our conception of the universe as a physical system or a set of interacting physical systems?

In the course of working on the series, certain other important themes emerged which could not be fully developed simply because of the limitations of the format. I want to make them fully explicit in this introduction, and in doing so I think I can help the reader to understand better the chapters which follow.

The first theme is how little we know of the functioning of the human brain, and how much the pretensions of certain theories depend on this ignorance. As David Hubel, the neurophysiologist, wrote in 1978: 'Our knowledge of the brain is in a very primitive state. While for some regions we have developed some kind of functional concept, there are others, the size of one's fist, of which it can almost be said that we are in the same state of knowledge as we were with regard to the heart before we realised that it pumped blood.' And indeed, if the interested layman picks up any of a half a dozen standard text books on the brain, as I did, and approaches them in an effort to get the answers to the sorts of questions that would immediately occur to any curious person, he is likely to be disappointed. What exactly is the neurophysiology of consciousness? Why do we need sleep? Why exactly does alcohol make us drunk? How exactly are memories stored in the brain? At the time of this writing, we simply do not know the answers to any of these fundamental questions. Many of the claims made about the mind in various disciplines ranging from Freudian psychology to artificial intelligence depend on this sort of ignorance. Such claims live in the holes in our knowledge.

On the traditional account of the brain, the account that takes the neuron as the fundamental unit of brain functioning, the most remarkable thing about brain functioning is simply this. All of the enormous variety of inputs that the brain receives – the photons that strike the retina, the sound waves that stimulate the ear drum, the pressure on the skin that activates nerve endings for pressure, heat, cold, and pain, etc. – all of these inputs are converted into one common medium: variable rates of neuron firing. Furthermore, and equally remarkably, these variable rates of neuron firing in different neuronal circuits and different local conditions in the brain produce all of the variety of our mental life. The smell of a rose, the experience of the blue of the sky, the taste of onions, the thought of a mathematical formula: all of these are produced by variable rates of neuron-firing, in different circuits, relative to different local conditions in the brain.

Now what exactly are these different neuronal circuits and what are the different local environments that account for the differences in our mental life? In detail no one knows, but we do have good evidence that certain regions of the brain are specialised for certain kinds of experiences. The visual cortex plays a special role in visual experiences, the auditory cortex in auditory experiences, etc. Suppose that auditory stimuli were fed to the visual cortex and visual stimuli were fed to the auditory cortex. What would happen? As far as I know, no one has ever done the experiment, but it seems reasonable to suppose that the auditory stimulus would be 'seen', that is, that it would produce visual experiences, and the visual stimulus would be 'heard', that is, it would produce auditory experiences, and both of these because of specific, though largely unknown, features of the visual and auditory cortex respectively. Though this hypothesis is speculative, it has some independent support if you reflect on the fact that a punch in the eye produces a visual flash ('seeing stars') even though it is not an optical stimulus.

A second theme that runs throughout these chapters is that we have an inherited cultural resistance to treating the conscious mind as a biological phenomenon like any other. This goes back to Descartes in the seventeenth century. Descartes divided the world into two kinds of substances: mental substances and physical substances. Physical substances were the proper domain of science and mental substances were the property of religion. Something of an acceptance of this division exists even to the present day. So, for example, consciousness and subjectivity are often regarded as unsuitable topics for science. And this reluctance to deal with consciousness and subjectivity is part of a persistent objectifying tendency. People think science must be about objectively observable phenomena. On occasions when I have lectured to audiences of biologists and neurophysiologists, I have found many of them very reluctant to treat the mind in general and consciousness in particular as a proper domain of scientific investigation.

A third theme that runs, subliminally, through these chapters is that the traditional terminology we have for discussing these problems is in various ways inadequate. Of the three terms that go to make up the title, *Minds, Brains and Science*, only the second is at all well defined. By 'mind' I just mean the sequences of thoughts, feelings and experiences, whether conscious or unconscious, that go to make up our mental life. But the use of the noun 'mind' is dangerously inhabited by the ghosts of old philosophical theories. It is very difficult to resist the idea that the mind is a kind of a thing, or at least an arena, or at least some kind of black box in which all of these mental processes occur.

The situation with the word 'science' is even worse. I would gladly do without this word if I could. 'Science' has become something of an honorific term, and all sorts of disciplines that are quite unlike physics and chemistry are eager to call themselves 'sciences'. A good rule of thumb to keep in mind is that anything that calls itself 'science' probably isn't – for example, Christian science, or military science, and possibly even cognitive science or social science. The word 'science' tends to suggest a lot of researchers in white coats waving test tubes and peering at instruments. To many minds it suggests an arcane infallibility. The rival picture I want to suggest is this: what we are all aiming at in intellectual disciplines is knowledge and understanding. There is only knowledge and understanding, whether we have it in mathematics, literary criticism, history, physics, or philosophy. Some disciplines are more systematic than others, and we might want to reserve the word 'science' for them.

I am indebted to a rather large number of students, colleagues, and friends for their help in the preparation of the Reith Lectures, both the broadcast and this published version. I especially want to thank Alan Code, Rejane Carrion, Stephen Davies, Hubert Dreyfus, Walter Freeman, Barbara Horan, Paul Kube, Karl Pribram, Gunther Stent, and Vanessa Whang.

The BBC was exceptionally helpful. George Fischer, the Head of the Talks Department, was very supportive; and my producer, Geoff Deehan, was simply excellent. My greatest debts are to my wife, Dagmar Searle, who assisted me at every step of the way, and to whom this book is dedicated.

# ONE
## THE MIND-BODY PROBLEM

For thousands of years, people have been trying to understand their relationship to the rest of the universe. For a variety of reasons many philosophers today are reluctant to tackle such big problems. Nonetheless, the problems remain, and in this book I am going to attack some of them.

At the moment, the biggest problem is this: We have a certain commonsense picture of ourselves as human beings which is very hard to square with our overall 'scientific' conception of the physical world. We think of ourselves as *conscious*, *free*, *mindful*, *rational* agents in a world that science tells us consists entirely of mindless, meaningless physical particles. Now, how can we square these two conceptions? How, for example, can it be the case that the world contains nothing but unconscious physical particles, and yet that it also contains consciousness? How can a mechanical universe contain intentionalistic human beings – that is, human beings that can represent the world to themselves? How, in short, can an essentially meaningless world contain meanings?

Such problems spill over into other more contemporary-sounding issues: How should we interpret recent work in computer science and artificial intelligence – work aimed at making intelligent machines? Specifically, does the digital computer give us the right picture of the human mind? And why is it that the social sciences in general have not given us insights into ourselves comparable to the insights that the natural sciences have given us into the rest of nature? What is the relation between the ordinary, commonsense explanations we accept of the way people behave and scientific modes of explanation?

In this first chapter, I want to plunge right into what many philosophers think of as the hardest problem of all: What is the relation of our minds to the rest of the universe? This, I am sure you will recognise, is the traditional mind-body or mind-brain problem. In its contemporary version it usually takes the form: how does the mind relate to the brain?

I believe that the mind-body problem has a rather simple solution, one that is consistent both with what we know about neurophysiology and with our commonsense conception of the nature of mental states – pains, beliefs, desires and so on. But before presenting that solution, I want to ask why the mind-body problem seems so intractable. Why do we still have in philosophy and psychology after all these centuries a 'mind-body problem' in a way that we do not have, say, a 'digestion-stomach problem'? Why does the mind seem more mysterious than other biological phenomena?

I am convinced that part of the difficulty is that we persist in talking about a twentieth-century problem in an outmoded seventeenth-century vocabulary. When I was an undergraduate, I remember being dissatisfied with the choices that were apparently available in the philosophy of mind: you could be either a monist or a dualist. If you were a monist, you could be either a materialist or an idealist. If you were a materialist, you could be either a behaviourist or a physicalist. And so on. One of my aims in what follows is to try to break out of these tired old categories. Notice that nobody feels he has to choose between monism and dualism where the 'digestion-stomach problem' is concerned. Why should it be any different with the 'mind-body problem'?

But, vocabulary apart, there is still a problem or family of problems. Since Descartes, the mind-body problem has taken the following form: how can we account for the relationships between two apparently completely different kinds of things? On the one hand, there are mental things, such as our thoughts and feelings; we think of them as subjective, conscious, and immaterial. On the other hand, there are physical things; we think of them as having mass, as extended in space, and as causally interacting with other physical things. Most attempted solutions to the mind-body problem wind up by denying the existence of, or in some way downgrading the status of, one or the other of these types of things. Given the successes of the physical sciences, it is not surprising that in our stage of intellectual development the temptation is to downgrade the status of mental entities. So, most of the recently fashionable materialist conceptions of the mind – such as behaviourism, functionalism, and physicalism – end up by denying, implicitly or explicitly, that there are any such things as minds as we ordinarily think of them. That is, they deny that we do really *intrinsically* have subjective, conscious, mental states and that they are as real and as irreducible as anything else in the universe.

Now, why do they do that? Why is it that so many theorists end up denying the intrinsically mental character of mental phenomena? If we can answer that question, I believe that we will understand why the mind-body problem has seemed so intractable for so long.

There are four features of mental phenomena which have made them seem impossible to fit into our 'scientific' conception of the world as made up of material things. And it is these four features that have made the mind-body problem really difficult. They are so embarrassing that they have led many thinkers in philosophy, psychology, and artificial intelligence to say strange and implausible things about the mind.

The most important of these features is consciousness. I, at the moment of writing this, and you, at the moment of reading it, are both conscious. It is just a plain fact about the world that it contains such conscious mental states and events, but it is hard to see how mere physical systems could have consciousness. How could such a thing occur? How, for example, could this grey and white gook inside my skull be conscious?

I think the existence of consciousness ought to seem amazing to us. It is easy enough to imagine a universe without it, but if you do, you will see that you have imagined a universe that is truly meaningless. Consciousness is the central fact of specifically human existence because without it all of the other specifically human aspects of our existence – language, love, humour, and so on – would be impossible. I believe it is, by the way, something of a scandal that contemporary discussions in philosophy and psychology have so little of interest to tell us about consciousness.

The second intractable feature of the mind is what philosophers and psychologists call 'intentionality', the feature by which our mental states are directed at, or about, or refer to, or are of objects and states of affairs in the world other than themselves. 'Intentionality', by the way, doesn't just refer to intentions, but also to beliefs, desires, hopes, fears, love, hate, lust, disgust, shame, pride, irritation, amusement, and all of those mental states (whether conscious or unconscious) that refer to, or are about, the world apart from the mind. Now the question about intentionality is much like the question about consciousness. How can this stuff inside my head be *about* anything? How can it *refer* to anything? After all, this stuff in the skull consists of 'atoms in the void', just as all of the rest of material reality consists of atoms in the void. Now how, to put it crudely, can atoms in the void represent anything?

The third feature of the mind that seems difficult to accommodate within a scientific conception of reality is the subjectivity of mental states. This subjectivity is marked by such facts as that I can feel my pains, and you can't. I see the world from my point of view; you see it from your point of view. I am aware of myself and my internal mental states, as quite distinct from the selves and mental states of other people. Since the seventeenth century we have come to think of reality as something which must be equally accessible to all competent observers – that is, we think it must be objective. Now, how are we to accommodate the reality of *subjective* mental phenomena with the scientific conception of reality as totally *objective*?

Finally, there is a fourth problem, the problem of mental causation. We all suppose, as part of common sense, that our thoughts and feelings make a real difference to the way we behave, that they actually have some *causal* effect on the physical world. I decide, for example, to raise my arm and – lo and behold – my arm goes up. But if our thoughts and feelings are truly mental, how can they affect anything physical? How could something mental make a physical difference? Are we supposed to think that our thoughts and feelings can somehow produce chemical effects on our brains and the rest of our nervous system? How could such a thing occur? Are we supposed to think that thoughts can wrap themselves around the axons or shake the dendrites or sneak inside the cell wall and attack the cell nucleus?

But unless some such connection takes place between the mind and the brain, aren't we just left with the view that the mind doesn't matter, that it is as unimportant causally as the froth on the wave is to the movement of the wave? I suppose if the froth were conscious, it might think to itself: 'What a tough job it is pulling these waves up on the beach and then pulling them out again, all day long!' But we know the froth doesn't make any important difference. Why do we suppose our mental life is any more important than a froth on the wave of physical reality?

These four features, consciousness, intentionality, subjectivity, and mental causation are what make the mind-body problem seem so difficult. Yet, I want to say, they are all real features of our mental lives. Not every mental state has all of them. But any satisfactory account of the mind and of mind-body relations must take account of all four features. If your theory ends up by denying any one of them, you know you must have made a mistake somewhere.

The first thesis I want to advance toward 'solving the mind-body problem' is this:

*Mental phenomena, all mental phenomena whether conscious or unconscious, visual or auditory, pains, tickles, itches, thoughts, indeed, all of our mental life, are caused by processes going on in the brain.*

To get a feel for how this works, let's try to describe the causal processes in some detail for at least one kind of mental state. For example, let's consider pains. Of course, anything we say now may seem wonderfully quaint in a generation, as our knowledge of how the brain works increases. Still, the *form* of the explanation can remain valid even though the *details* are altered. On current views, pain signals are transmitted from sensory nerve endings to the spinal cord by at least two types of fibres – there are Delta A fibres, which are specialised for prickling sensations, and C fibres, which are specialised for burning and aching sensations. In the spinal cord, they pass through a region called the tract of Lissauer and terminate on the neurons of the cord. As the signals go up the spine, they enter the brain by two separate pathways: the prickling pain pathway and the burning pain pathway. Both pathways go through the thalamus, but the prickling pain is more localised afterwards in the somato-sensory cortex, whereas the burning pain pathway transmits signals, not only upwards into the cortex, but also laterally into the hypothalamus and other regions at the base of the brain. Because of these differences, it is much easier for us to localise a prickling sensation – we can tell fairly accurately where someone is sticking a pin into our skin, for example – whereas burning and aching pains can be more distressing because they activate more of the nervous system. The actual sensation of pain appears to be caused both by the stimulation of the basal regions of the brain, especially the thalamus, and the stimulation of the somato-sensory cortex.

Now for the purposes of this discussion, the point we need to hammer home is this: our sensations of pains are caused by a series of events that begin at free nerve endings and end in the thalamus and in other regions of the brain. Indeed, as far as the actual sensations are concerned, the events inside the central nervous system are quite sufficient to cause pains – we know this both from the phantom-limb pains felt by amputees and the pains caused by artificially stimulating relevant portions of the brain. I want to suggest that what is true of pain is true of mental phenomena generally. To put it crudely, and counting all of the central nervous system as part of the brain for our present discussion, everything that matters for our mental life, all of our thoughts and feelings, are caused by processes inside the brain. As far as causing mental states is concerned, the crucial step is the one that goes on inside the head, not the external or peripheral stimulus. And the argument for this is simple. If the events outside the central nervous system occurred, but nothing happened in the brain, there would be no mental events. But if the right things happened in the brain, the mental events would occur even if there was no outside stimulus. (And that, by the way, is the principle on which surgical anaesthesia works: the outside stimulus is prevented from having the relevant effects on the central nervous system.)

But if pains and other mental phenomena are caused by processes in the brain, one wants to know: what are pains? What are they really? Well, in the case of pains, the obvious answer is that they are unpleasant sorts of sensations. But that answer leaves us unsatisfied because it doesn't tell us how pains fit into our overall conception of the world.

Once again, I think the answer to the question is obvious, but it will take some spelling out. To our first claim – that pains and other mental phenomena are caused by brain processes, we need to add a second claim:

*Pains and other mental phenomena just are features of the brain (and perhaps the rest of the central nervous system).*

One of the primary aims of this chapter is to show how *both* of these propositions can be true together. How can it be both the case that brains cause minds and yet minds just are features of brains? I believe it is the failure to see how both these propositions can be true together that has blocked a solution to the mind-body problem for so long. There are different levels of confusion that such a pair of ideas can generate. If mental and physical phenomena have cause and effect relationships, how can one be a feature of the other? Wouldn't that imply that the mind caused itself – the dreaded doctrine of *causa sui*? But at the bottom of our puzzlement is a misunderstanding of causation. It is tempting to think that whenever A causes B there must be two discrete events, one identified as the cause, the other identified as the effect; that all causation functions in the same way as billiard balls hitting each other. This crude model of the causal relationships between the brain and the mind inclines us to accept some kind of dualism; we are inclined to think that events in one material realm, the 'physical', cause events in another insubstantial realm, the 'mental'. But that seems to me a mistake. And the way to remove the mistake is to get a more sophisticated concept of causation. To do this, I will turn away from the relations between mind and brain for a moment to observe some other sorts of causal relationships in nature.

A common distinction in physics is between micro- and macro-properties of systems – the small and large scales. Consider, for example, the desk at which I am now sitting, or the glass of water in front of me. Each object is composed of micro-particles. The micro-particles have features at the level of molecules and atoms as well as at the deeper level of subatomic particles. But each object also has certain properties such as the solidity of the table, the liquidity of the water, and the transparency of the glass, which are surface or global features of the physical systems. Many such surface or global properties can be causally explained by the behaviour of elements at the micro-level. For example, the solidity of the table in front of me is explained by the lattice structure occupied by the molecules of which the table is composed. Similarly, the liquidity of the water is explained by the nature of the interactions between the $H_2O$ molecules. Those macro-features are causally explained by the behaviour of elements at the micro-level.

I want to suggest that this provides a perfectly ordinary model for explaining the puzzling relationships between the mind and the brain. In the case of liquidity, solidity, and transparency, we have no difficulty at all in supposing that the surface features are *caused by* the behaviour of elements at the micro-level, and at the same time we accept that the surface phenomena *just are* features of the very systems in question. I think the clearest way of stating this point is to say that the surface feature is both *caused by* the behaviour of micro-elements, and at the same time is *realised in* the system that is made up of the micro-elements. There is a cause and effect relationship, but at the same time the surface features are just higher level features of the very system whose behaviour at the micro-level causes those features.

In objecting to this someone might say that liquidity, solidity, and so on are identical with features of the micro-structure. So, for example, we might just define solidity as the lattice structure of the molecular arrangement, just as heat often is identified with the mean kinetic energy of molecule movements. This point seems to me correct but not really an objection to the analysis that I am proposing. It is a characteristic of the progress of science that an expression that is originally defined in terms of surface features, features accessible to the senses, is subsequently defined in terms of the micro-structure that causes the surface features. Thus, to take the example of solidity, the table in front of me is solid in the ordinary sense that it is rigid, it resists pressure, it supports books, it is not easily penetrable by most other objects such as other tables, and so on. Such is the commonsense notion of solidity. And in a scientific vein one can define solidity as whatever micro-structure causes these gross observable features. So one can then say either that solidity just is the lattice structure of the system of molecules and that solidity so defined causes, for example, resistance to touch and pressure. Or one can say that solidity consists of such high level features as rigidity and resistance to touch and pressure and that it is caused by the behaviour of elements at the micro-level.

If we apply these lessons to the study of the mind, it seems to me that there is no difficulty in accounting for the relations of the mind to the brain in terms of the brain's functioning to cause mental states. Just as the liquidity of the water is caused by the behaviour of elements at the micro-level, and yet at the same time it is a feature realised in the system of micro-elements, so in exactly that sense of 'caused by' and 'realised in' mental phenomena are caused by processes going on in the brain at the neuronal or modular level, and at the same time they are realised in the very system that consists of neurons. And just as we need the micro/macro distinction for any physical system, so for the same reasons we need the micro/macro distinction for the brain. And though we can say of a system of particles that it is 10°C or it is solid or it is liquid, we cannot say of any given particle that this particle is solid, this particle is liquid, this particle is 10°C. I can't for example reach into this glass of water, pull out a molecule and say: 'This one's wet'.

In exactly the same way, as far as we know anything at all about it, though we can say of a particular brain: 'This brain is conscious', or: 'This brain is experiencing thirst or pain', we can't say of any particular neuron in the brain: 'This neuron is in pain, this neuron is experiencing thirst'. To repeat this point, though there are enormous empirical mysteries about how the brain works in detail, there are no logical or philosophical or metaphysical obstacles to accounting for the relation between the mind and the brain in terms that are quite familiar to us from the rest of nature. Nothing is more common in nature than for surface features of a phenomenon to be both caused by and realised in a micro-structure, and those are exactly the relationships that are exhibited by the relation of mind to brain.

Let us now return to the four problems that I said faced any attempt to solve the mind-brain problem.

First, how is consciousness possible?

The best way to show how something is possible is to show how it actually exists. We have already given a sketch of how pains are actually caused by neurophysiological processes going on in the thalamus and the sensory cortex. Why is it then that many people feel dissatisfied with this sort of answer? I think that by pursuing an analogy with an earlier problem in the history of science we can dispel this sense of puzzlement. For a long time many biologists and philosophers thought it was impossible, in principle, to account for the existence of *life* on purely biological grounds. They thought that in addition to the biological processes some other element must be necessary, some *élan vital* must be postulated in order to lend life to what was otherwise dead and inert matter. It is hard today to realise how intense the dispute was between vitalism and mechanism even a generation ago, but today these issues are no longer taken seriously. Why not? I think it is not so much because mechanism won and vitalism lost, but because we have come to understand better the biological character of the processes that are characteristic of living organisms. Once we understand how the features that are characteristic of living beings have a biological explanation, it no longer seems mysterious to us that matter should be alive. I think that exactly similar considerations should apply to our discussions of consciousness. It should seem no more mysterious, in principle, that this hunk of matter, this grey and white oatmeal-textured substance of the brain, should be conscious than it seems mysterious that this other hunk of matter, this collection of nucleo-protein molecules stuck onto a calcium frame, should be alive. The way, in short, to dispel the mystery is to understand the processes. We do not yet fully understand the processes, but we understand their general *character*, we understand that there are certain specific electrochemical activities going on among neurons or neuron-modules and perhaps other features of the brain and these processes cause consciousness.

Our second problem was, how can atoms in the void have intentionality? How can they be about something?

As with our first question, the best way to show how something is possible is to show how it actually exists. So let's consider thirst. As far as we know anything about it, at least certain kinds of thirst are caused in the hypothalamus by sequences of nerve firings. These firings are in turn caused by the action of angiotensin in the hypothalamus, and angiotensin, in turn, is synthesised by renin, which is secreted by the kidneys. Thirst, at least of these kinds, is caused by a series of events in the central nervous system, principally the hypothalamus, and it is realised in the hypothalamus. To be thirsty is to have, among other things, the desire to drink. Thirst is therefore an intentional state: it has content; its content determines under what conditions it is satisfied, and it has all the rest of the features that are common to intentional states.

As with the 'mysteries' of life and consciousness, the way to master the mystery of intentionality is to describe in as much detail as we can how the phenomena are caused by biological processes while being at the same time realised in biological systems. Visual and auditory experiences, tactile sensations, hunger, thirst, and sexual desire, are all caused by brain processes and they are realised in the structure of the brain, and they are all intentional phenomena.

I am not saying we should lose our sense of the mysteries of nature. On the contrary, the examples I have cited are all in a sense astounding. But I am saying that they are neither more nor less mysterious than other astounding features of the world, such as the existence of gravitational attraction, the process of photosynthesis, or the size of the Milky Way.

Our third problem: how do we accommodate the subjectivity of mental states within an objective conception of the real world?

It seems to me a mistake to suppose that the definition of reality should exclude subjectivity. If 'science' is the name of the collection of objective and systematic truths we can state about the world, then the existence of subjectivity is an objective scientific fact like any other. If a scientific account of the world attempts to describe how things are, then one of the features of the account will be the subjectivity of mental states, since it is just a plain fact about biological evolution that it has produced certain sorts of biological systems, namely human and certain animal brains, that have subjective features. My present state of consciousness is a feature of my brain, but its conscious aspects are accessible to me in a way that they are not accessible to you. And your present state of consciousness is a feature of your brain and its conscious aspects are accessible to you in a way that they are not accessible to me. Thus the existence of subjectivity is an objective fact of biology. It is a persistent mistake to try to define 'science' in terms of certain features of existing scientific theories. But once this provincialism is perceived to be the prejudice it is, then any domain of facts whatever is a subject of systematic investigation. So, for example, if God existed, then that fact would be a fact like any other. I do not know whether God exists, but I have no doubt at all that subjective mental states exist, because I am now in one and so are you. If the fact of subjectivity runs counter to a certain definition of 'science', then it is the definition and not the fact which we will have to abandon.

Fourth, the problem of mental causation for our present purpose is to explain how mental events can cause physical events. How, for example, could anything as 'weightless' and 'ethereal' as a thought give rise to an action?

The answer is that thoughts are not weightless and ethereal. When you have a thought, brain activity is actually going on.

Brain activity causes bodily movements by physiological processes. Now, because mental states are features of the brain, they have two levels of description – a higher level in mental terms, and a lower level in physiological terms. The very same causal powers of the system can be described at either level.

Once again, we can use an analogy from physics to illustrate these relationships. Consider hammering a nail with a hammer. Both hammer and nail have a certain kind of solidity. Hammers made of cottonwool or butter will be quite useless, and hammers made of water or steam are not hammers at all. Solidity is a real causal property of the hammer. But the solidity itself is caused by the behaviour of particles at the micro-level and it is realised in the system which consists of micro-elements. The existence of two causally real levels of description in the brain, one a macro-level of mental processes and the other a micro-level of neuronal processes is exactly analogous to the existence of two causally real levels of description of the hammer. Consciousness, for example, is a real property of the brain that can cause things to happen. My conscious attempt to perform an action such as raising my arm causes the movement of the arm. At the higher level of description, the intention to raise my arm causes the movement of the arm. But at the lower level of description, a series of neuron firings starts a chain of events that results in the contraction of the muscles. As with the case of hammering a nail, the same sequence of events has two levels of description. Both of them are causally real, and the higher level causal features are both caused by and realised in the structure of the lower level elements.

To summarise: on my view, the mind and the body interact, but they are not two different things, since mental phenomena just are features of the brain. One way to characterise this position is to see it as an assertion of both physicalism and mentalism. Suppose we define 'naive physicalism' to be the view that all that exists in the world are physical particles with their properties and relations. The power of the physical model of reality is so great that it is hard to see how we can seriously challenge naive physicalism. And let us define 'naive mentalism' to be the view that mental phenomena really exist. There really are mental states; some of them are conscious; many have intentionality; they all have subjectivity; and many of them function causally in determining physical events in the world. The thesis of this first chapter can now be stated quite simply. Naive mentalism and naive physicalism are perfectly consistent with each other. Indeed, as far as we know anything about how the world works, they are not only consistent, they are both true.

# TWO

## CAN COMPUTERS THINK?

In the previous chapter, I provided at least the outlines of a solution to the so-called 'mind-body problem'. Though we do not know in detail how the brain functions, we do know enough to have an idea of the general relationships between brain processes and mental processes. Mental processes are caused by the behaviour of elements of the brain. At the same time, they are realised in the structure that is made up of those elements. I think this answer is consistent with the standard biological approaches to biological phenomena. Indeed, it is a kind of commonsense answer to the question, given what we know about how the world works. However, it is very much a minority point of view. The prevailing view in philosophy, psychology, and artificial intelligence is one which emphasises the analogies between the functioning of the human brain and the functioning of digital computers. According to the most extreme version of this view, the brain is just a digital computer and the mind is just a computer program. One could summarise this view – I call it 'strong artificial intelligence', or 'strong AI' – by saying that the mind is to the brain, as the program is to the computer hardware.

This view has the consequence that there is nothing essentially biological about the human mind. The brain just happens to be one of an indefinitely large number of different kinds of hardware computers that could sustain the programs which make up human intelligence. On this view, any physical system whatever that had the right program with the right inputs and outputs would have a mind in exactly the same sense that you and I have minds. So, for example, if you made a computer out of old beer cans powered by windmills; if it had the right program, it would have to have a mind. And the point is not that for all we know it might have thoughts and feelings, but rather that it must have thoughts and feelings, because that is all there is to having thoughts and feelings: implementing the right program.

Most people who hold this view think we have not yet designed programs which are minds. But there is pretty much general agreement among them that it's only a matter of time until computer scientists and workers in artificial intelligence design the appropriate hardware and programs which will be the equivalent of human brains and minds. These will be artificial brains and minds which are in every way the equivalent of human brains and minds.

Many people outside of the field of artificial intelligence are quite amazed to discover that anybody could believe such a view as this. So, before criticising it, let me give you a few examples of the things that people in this field have actually said. Herbert Simon of Carnegie-Mellon University says that we already have machines that can literally think. There is no question of waiting for some future machine, because existing digital computers already have thoughts in exactly the same sense that you and I do. Well, fancy that! Philosophers have been worried for centuries about whether or not a machine could think, and now we discover that they already have such machines at Carnegie-Mellon. Simon's colleague Alan Newell claims that we have now discovered (and notice that Newell says 'discovered' and not 'hypothesised' or 'considered the possibility', but we have *discovered*) that intelligence is just a matter of physical symbol manipulation; it has no essential connection with any specific kind of biological or physical wetware or hardware. Rather, any system whatever that is capable of manipulating physical symbols in the right way is capable of intelligence in the same literal sense as human intelligence of human beings. Both Simon and Newell, to their credit, emphasise that there is nothing metaphorical about these claims; they mean them quite literally. Freeman

Dyson is quoted as having said that computers have an advantage over the rest of us when it comes to evolution. Since consciousness is just a matter of formal processes, in computers these formal processes can go on in substances that are much better able to survive in a universe that is cooling off than beings like ourselves made of our wet and messy materials. Marvin Minsky of MIT says that the next generation of computers will be so intelligent that we will 'be lucky if they are willing to keep us around the house as household pets'. My all-time favourite in the literature of exaggerated claims on behalf of the digital computer is from John McCarthy, the inventor of the term 'artificial intelligence'. McCarthy says even 'machines as simple as thermostats can be said to have beliefs'. And indeed, according to him, almost any machine capable of problem-solving can be said to have beliefs. I admire McCarthy's courage. I once asked him: 'What beliefs does your thermostat have?' And he said: 'My thermostat has three beliefs – it's too hot in here, it's too cold in here, and it's just right in here.' As a philosopher, I like all these claims for a simple reason. Unlike most philosophical theses, they are reasonably clear, and they admit of a simple and decisive refutation. It is this refutation that I am going to undertake in this chapter.

The nature of the refutation has nothing whatever to do with any particular stage of computer technology. It is important to emphasise this point because the temptation is always to think that the solution to our problems must wait on some as yet uncreated technological wonder. But in fact, the nature of the refutation is completely independent of any state of technology. It has to do with the very definition of a digital computer, with what a digital computer is.

It is essential to our conception of a digital computer that its operations can be specified purely formally; that is, we specify the steps in the operation of the computer in terms of abstract symbols – sequences of zeroes and ones printed on a tape, for example. A typical computer 'rule' will determine that when a machine is in a certain state and it has a certain symbol on its tape, then it will perform a certain operation such as erasing the symbol or printing another symbol and then enter another state such as moving the tape one square to the left. But the symbols have no meaning; they have no semantic content; they are not about anything. They have to be specified purely in terms of their formal or syntactical structure. The zeroes and ones, for example, are just numerals; they don't even stand for numbers. Indeed, it is this feature of digital computers that makes them so powerful. One and the same type of hardware, if it is appropriately designed, can be used to run an indefinite range of different programs. And one and the same program can be run on an indefinite range of different types of hardwares.

But this feature of programs, that they are defined purely formally or syntactically, is fatal to the view that mental processes and program processes are identical. And the reason can be stated quite simply. There is more to having a mind than having formal or syntactical processes. Our internal mental states, by definition, have certain sorts of contents. If I am thinking about Kansas City or wishing that I had a cold beer to drink or wondering if there will be a fall in interest rates, in each case my mental state has a certain mental content in addition to whatever formal features it might have. That is, even if my thoughts occur to me in strings of symbols, there must be more to the thought than the abstract strings, because strings by themselves can't have any meaning. If my thoughts are to be *about* anything, then the strings must have a *meaning* which makes the thoughts about those things. In a word, the mind has more than a syntax, it has a semantics. The reason that no computer program can ever be a mind is simply that a computer program is only syntactical, and minds are more than syntactical. Minds are semantical, in the sense that they have more than a formal structure, they have a content.

To illustrate this point I have designed a certain thought-experiment. Imagine that a bunch of computer programmers have written a program that will enable a computer to simulate the understanding of Chinese. So, for example, if the computer is given a question in Chinese, it will match the question against its memory, or data base, and produce appropriate answers to the questions in Chinese. Suppose for the sake of argument that the computer's answers are as good as those of a native Chinese speaker. Now then, does the computer, on the basis of this, understand Chinese, does it literally understand Chinese, in the way that Chinese speakers understand Chinese? Well, imagine that you are locked in a room, and in this room are several baskets full of Chinese symbols. Imagine that you (like me) do not understand a word of Chinese, but that you are given a rule book in English for manipulating these Chinese symbols. The rules specify the manipulations of the symbols purely formally, in terms of their syntax, not their semantics. So the rule might say: 'Take a squiggle-squiggle sign out of basket number one and put it next to a squoggle-squoggle sign from basket number two.' Now suppose that some other Chinese symbols are passed into the room, and that you are given further rules for passing back Chinese symbols out of the room. Suppose that unknown to you the symbols passed into the room are called 'questions' by the people outside the room, and the symbols you pass back out of the room are called 'answers to the questions'. Suppose, furthermore, that the programmers are so good at designing the programs and that you are so good at manipulating the symbols, that very soon your answers are indistinguishable from those of a native Chinese speaker. There you are locked in your room shuffling your Chinese symbols and passing out Chinese symbols in response to incoming Chinese symbols. On the basis of the situation as I have described it, there is no way you could learn any Chinese simply by manipulating these formal symbols.

Now the point of the story is simply this: by virtue of implementing a formal computer program from the point of view of an outside observer, you behave exactly as if you understood Chinese, but all the same you don't understand a word of Chinese. But if going through the appropriate computer program for understanding Chinese is not enough to give *you* an understanding of Chinese, then it is not enough to give *any other digital computer* an understanding of Chinese. And again, the reason for this can be stated quite simply. If you don't understand Chinese, then no other computer could understand Chinese because no digital computer, just by virtue of running a program, has anything that you don't have. All that the computer has, as you have, is a formal program for manipulating uninterpreted Chinese symbols. To repeat, a computer has a syntax, but no semantics. The whole point of the parable of the Chinese room is to remind us of a fact that we knew all along. Understanding a language, or indeed, having mental states at all, involves more than just having a bunch of formal symbols. It involves having an interpretation, or a meaning attached to those symbols. And a digital computer, as defined, cannot have more than just formal symbols because the operation of the computer, as I said earlier, is defined in terms of its ability to implement programs. And these programs are purely formally specifiable – that is, they have no semantic content.

We can see the force of this argument if we contrast what it is like to be asked and to answer questions in English, and to be asked and to answer questions in some language where we have no knowledge of any of the meanings of the words. Imagine that in the Chinese room you are also given questions in English about such things as your age or your life history, and that you answer these questions. What is the difference between the Chinese case and the English case? Well again, if like me you understand no Chinese and you do understand English, then the difference is obvious. You understand the questions in English because they are expressed in symbols whose meanings are known to you. Similarly, when you give the answers in English you are producing symbols which are meaningful to you. But in the case of the Chinese, you have none of that. In the case of the Chinese, you simply manipulate formal symbols according to a computer program, and you attach no meaning to any of the elements.

Various replies have been suggested to this argument by workers in artificial intelligence and in psychology, as well as philosophy. They all have something in common; they are all inadequate. And there is an obvious reason why they have to be inadequate, since the argument rests on a very simple logical truth, namely, syntax alone is not sufficient for semantics, and digital computers insofar as they are computers have, by definition, a syntax alone.

I want to make this clear by considering a couple of the arguments that are often presented against me.

Some people attempt to answer the Chinese room example by saying that the whole system understands Chinese. The idea here is that though I, the person in the room manipulating the symbols do not understand Chinese, I am just the central processing unit of the computer system. They argue that it is the whole system, including the room, the baskets full of symbols and the ledgers containing the programs and perhaps other items as well, taken as a totality, that understands Chinese. But this is subject to exactly the same objection I made before. There is no way that the system can get from the syntax to the semantics. I, as the central processing unit have no way of figuring out what any of these symbols means; but then neither does the whole system.

Another common response is to imagine that we put the Chinese understanding program inside a robot. If the robot moved around and interacted causally with the world, wouldn't that be enough to guarantee that it understood Chinese? Once again the inexorability of the semantics-syntax distinction overcomes this manoeuvre. As long as we suppose that the robot has only a computer for a brain then, even though it might behave exactly as if it understood Chinese, it would still have no way of getting from the syntax to the semantics of Chinese. You can see this if you imagine that I am the computer. Inside a room in the robot's skull I shuffle symbols without knowing that some of them come in to me from television cameras attached to the robot's head and others go out to move the robot's arms and legs. As long as all I have is a formal computer program, I have no way of attaching any meaning to any of the symbols. And the fact that the robot is engaged in causal interactions with the outside world won't help me to attach any meaning to the symbols unless I have some way of finding out about that fact. Suppose the robot picks up a hamburger and this triggers the symbol for hamburger to come into the room. As long as all I have is the symbol with no knowledge of its causes or how it got there, I have no way of knowing what it means. The causal interactions between the robot and the rest of the world are irrelevant unless those causal interactions are represented in some mind or other. But there is no way they can be if all that the so-called mind consists of is a set of purely formal, syntactical operations.

It is important to see exactly what is claimed and what is not claimed by my argument. Suppose we ask the question that I mentioned at the beginning: 'Could a machine think?' Well, in one sense, of course, we are all machines. We can construe the stuff inside our heads as a meat machine. And of course, we can all think. So, in one sense of 'machine', namely that sense in which a machine is just a physical system which is capable of performing certain kinds of operations, in that sense, we are all machines, and we can think. So, trivially, there are machines that can think. But that wasn't the question that bothered us. So let's try a different formulation of it. Could an artefact think? Could a man-made machine think? Well, once again, it depends on the kind of artefact. Suppose we designed a machine that was molecule-for-molecule indistinguishable from a human being. Well then, if you can duplicate the causes, you can presumably duplicate the effects. So once again, the answer to that question is, in principle at least, trivially yes. If you could build a machine that had the same structure as a human being, then presumably that machine would be able to think. Indeed, it would be a surrogate human being. Well, let's try again.

The question isn't: 'Can a machine think?' or: 'Can an artefact think?' The question is: 'Can a digital computer think?' But once again we have to be very careful in how we interpret the question. From a mathematical point of view, anything whatever can be described *as if* it were a digital computer. And that's because it can be described as instantiating or implementing a computer program. In an utterly trivial sense, the pen that is on the desk in front of me can be described as a digital computer. It just happens to have a very boring computer program. The program says: 'Stay there.' Now since in this sense, anything whatever is a digital computer, because anything whatever can be described as implementing a computer program, then once again, our question gets a trivial answer. Of course our brains are digital computers, since they implement any number of computer programs. And of course our brains can think. So once again, there is a trivial answer to the question. But that wasn't really the question we were trying to ask. The question we wanted to ask is this: 'Can a digital computer, as defined, think?' That is to say: 'Is instantiating or implementing the right computer program with the right inputs and outputs, sufficient for, or constitutive of, thinking?' And to this question, unlike its predecessors, the answer is clearly 'no'. And it is 'no' for the reason that we have spelled out, namely, the computer program is defined purely syntactically. But thinking is more than just a matter of manipulating meaningless symbols, it involves meaningful semantic contents. These semantic contents are what we mean by 'meaning'.

It is important to emphasise again that we are not talking about a particular stage of computer technology. The argument has nothing to do with the forthcoming, amazing advances in computer science. It has nothing to do with the distinction between serial and parallel processes, or with the size of programs, or the speed of computer operations, or with computers that can interact causally with their environment, or even with the invention of robots. Technological progress is always grossly exaggerated, but even subtracting the exaggeration, the development of computers has been quite remarkable, and we can reasonably expect that even more remarkable progress will be made in the future. No doubt we will be much better able to simulate human behaviour on computers than we can at present, and certainly much better than we have been able to in the past. The point I am making is that if we are talking about having mental states, having a mind, all of these simulations are simply irrelevant. It doesn't matter how good the technology is, or how rapid the calculations made by the computer are. If it really is a computer, its operations have to be defined syntactically, whereas consciousness, thoughts, feelings, emotions, and all the rest of it involve more than a syntax. Those features, by definition, the computer is unable to *duplicate* however powerful may be its ability to *simulate*. The key distinction here is between duplication and simulation. And no simulation by itself ever constitutes duplication.

What I have done so far is give a basis to the sense that those citations I began this talk with are really as preposterous as they seem. There is a puzzling question in this discussion though, and that is: 'Why would anybody ever have thought that computers could think or have feelings and emotions and all the rest of it?' After all, we can do computer simulations of any process whatever that can be given a formal description. So, we can do a computer simulation of the flow of money in the British economy, or the pattern of power distribution in the Labour party. We can do computer simulation of rain storms in the home counties, or warehouse fires in East London. Now, in each of these cases, nobody supposes that the computer simulation is actually the real thing; no one supposes that a computer simulation of a storm will leave us all wet, or a computer simulation of a fire is likely to burn the house down. Why on earth would anyone in his right mind suppose a computer simulation of mental processes actually had mental processes? I don't really know the answer to that, since the idea seems to me, to put it frankly, quite crazy from the start. But I can make a couple of speculations.

First of all, where the mind is concerned, a lot of people are still tempted to some sort of behaviourism. They think if a system behaves as if it understood Chinese, then it really must understand Chinese. But we have already refuted this form of behaviourism with the Chinese room argument. Another assumption made by many people is that the mind is not a part of the biological world, it is not a part of the world of nature. The strong artificial intelligence view relies on that in its conception that the mind is purely formal; that somehow or other, it cannot be treated as a concrete product of biological processes like any other biological product. There is in these discussions, in short, a kind of residual dualism. AI partisans believe that the mind is more than a part of the natural biological world; they believe that the mind is purely formally specifiable. The paradox of this is that the AI literature is filled with fulminations against some view called 'dualism', but in fact, the whole thesis of strong AI rests on a kind of dualism. It rests on a rejection of the idea that the mind is just a natural biological phenomenon in the world like any other.

I want to conclude this chapter by putting together the thesis of the last chapter and the thesis of this one. Both of these theses can be stated very simply. And indeed, I am going to state them with perhaps excessive crudeness. But if we put them together I think we get a quite powerful conception of the relations of minds, brains and computers. And the argument has a very simple logical structure, so you can see whether it is valid or invalid. The first premise is:

1. *Brains cause minds.*

Now, of course, that is really too crude. What we mean by that is that mental processes that we consider to constitute a mind are caused, entirely caused, by processes going on inside the brain. But let's be crude, let's just abbreviate that as three words – brains cause minds. And that is just a fact about how the world works. Now let's write proposition number two:

2. *Syntax is not sufficient for semantics.*

That proposition is a conceptual truth. It just articulates our distinction between the notion of what is purely formal and what has content. Now, to these two propositions – that brains cause minds and that syntax is not sufficient for semantics – let's add a third and a fourth:

3. *Computer programs are entirely defined by their formal, or syntactical, structure.*

That proposition, I take it, is true by definition; it is part of what we mean by the notion of a computer program.

4. *Minds have mental contents; specifically, they have semantic contents.*

And that, I take it, is just an obvious fact about how our minds work. My thoughts, and beliefs, and desires are about something, or they refer to something, or they concern states of affairs in the world; and they do that because their content directs them at these states of affairs in the world. Now, from these four premises, we can draw our first conclusion; and it follows obviously from premises 2, 3 and 4:

CONCLUSION 1. *No computer program by itself is sufficient to give a system a mind. Programs, in short, are not minds, and they are not by themselves sufficient for having minds.*

Now, that is a very powerful conclusion, because it means that the project of trying to create minds solely by designing programs is doomed from the start. And it is important to re-emphasise that this has nothing to do with any particular state of technology or any particular state of the complexity of the program. This is a purely formal, or logical, result from a set of axioms which are agreed to by all (or nearly all) of the disputants concerned. That is, even most of the hardcore enthusiasts for artificial intelligence agree that in fact, as a matter of biology, brain processes cause mental states, and they agree that programs are defined purely formally. But if you put these conclusions together with certain other things that we know, then it follows immediately that the project of strong AI is incapable of fulfilment.

However, once we have got these axioms, let's see what else we can derive. Here is a second conclusion:

CONCLUSION 2. *The way that brain functions cause minds cannot be solely in virtue of running a computer program.*

And this second conclusion follows from conjoining the first premise together with our first conclusion. That is, from the fact that brains cause minds and that programs are not enough to do the job, it follows that the way that brains cause minds can't be solely by running a computer program. Now that also I think is an important result, because it has the consequence that the brain is not, or at least is not just, a digital computer. We saw earlier that anything can trivially be described as if it were a digital computer, and brains are no exception. But the importance of this conclusion is that the computational properties of the brain are simply not enough to explain its functioning to produce mental states. And indeed, that ought to seem a commonsense scientific conclusion to us anyway because all it does is remind us of the fact that brains are biological engines; their biology matters. It is not, as several people in artificial intelligence have claimed, just an irrelevant fact about the mind that it happens to be realised in human brains.

Now, from our first premise, we can also derive a third conclusion:

CONCLUSION 3. *Anything else that caused minds would have to have causal powers at least equivalent to those of the brain.*

And this third conclusion is a trivial consequence of our first premise. It is a bit like saying that if my petrol engine drives my car at seventy-five miles an hour, then any diesel engine that was capable of doing that would have to have a power output at least equivalent to that of my petrol engine. Of course, some other system might cause mental processes using entirely different chemical or biochemical features from those the brain in fact uses. It might turn out that there are beings on other planets, or in other solar systems, that have mental states and use an entirely different biochemistry from ours. Suppose that Martians arrived on earth and we concluded that they had mental states. But suppose that when their heads were opened up, it was discovered that all they had inside was green slime. Well still, the green slime, if it functioned to produce consciousness and all the rest of their mental life, would have to have causal powers equal to those of the human brain. But now, from our first conclusion, that programs are not enough, and our third conclusion, that any other system would have to have causal powers equal to the brain, conclusion four follows immediately:

CONCLUSION 4. *For any artefact that we might build which had mental states equivalent to human mental states, the implementation of a computer program would not by itself be sufficient. Rather the artefact would have to have powers equivalent to the powers of the human brain.*

The upshot of this discussion I believe is to remind us of something that we have known all along: namely, mental states are biological phenomena. Consciousness, intentionality, subjectivity and mental causation are all a part of our biological life history, along with growth, reproduction, the secretion of bile, and digestion.

# THREE
## COGNITIVE SCIENCE

We feel perfectly confident in saying things like: 'Basil voted for the Tories because he liked Mrs Thatcher's handling of the Falklands affair.' But we have no idea how to go about saying things like: 'Basil voted for the Tories because of a condition of his hypothalamus.' That is, we have commonsense explanations of people's behaviour in mental terms, in terms of their desires, wishes, fears, hopes, and so on. And we suppose that there must also be a neurophysiological sort of explanation of people's behaviour in terms of processes in their brains. The trouble is that the first of these sorts of explanations works well enough in practice, but is not scientific; whereas the second is certainly scientific, but we have no idea how to make it work in practice.

Now that leaves us apparently with a gap, a gap between the brain and the mind. And some of the greatest intellectual efforts of the twentieth century have been attempts to fill this gap, to get a science of human behaviour which was not just commonsense grandmother psychology, but was not scientific neurophysiology either. Up to the present time, without exception, the gap-filling efforts have been failures. Behaviourism was the most spectacular failure, but in my lifetime I have lived through exaggerated claims made on behalf of and eventually disappointed by games theory, cybernetics, information theory, structuralism, sociobiology, and a bunch of others. To anticipate a bit, I am going to claim that all the gap-filling efforts fail because there isn't any gap to fill.

The most recent gap-filling efforts rely on analogies between human beings and digital computers. On the most extreme version of this view, which I call 'strong artificial intelligence' or just 'strong AI', the brain is a digital computer and the mind is just a computer program. Now, that's the view I refuted in the last chapter. A related recent attempt to fill the gap is often called 'cognitivism', because it derives from work in cognitive psychology and artificial intelligence, and it forms the mainstream of a new discipline of 'cognitive science'. Like strong AI, it sees the computer as the right picture of the mind, and not just as a metaphor. But unlike strong AI, it does not, or at least it doesn't have to, claim that computers literally have thoughts and feelings.

If one had to summarise the research program of cognitivism it would look like this: Thinking is processing information, but information processing is just symbol manipulation. Computers do symbol manipulation. So the best way to study thinking (or as they prefer to call it, 'cognition') is to study computational symbol-manipulating programs, whether they are in computers or in brains. On this view, then, the task of cognitive science is to characterise the brain, not at the level of nerve cells, nor at the level of conscious mental states, but rather at the level of its functioning as an information processing system. And that's where the gap gets filled.

I cannot exaggerate the extent to which this research project has seemed to constitute a major breakthrough in the science of the mind. Indeed, according to its supporters, it might even be *the* breakthrough that will at last place psychology on a secure scientific footing now that it has freed itself from the delusions of behaviourism.

I am going to attack cognitivism in this lecture, but I want to begin by illustrating its attractiveness. We know that there is a level of naive, commonsense, grandmother psychology and also a level of neurophysiology – the level of neurons and neuron modules and synapses and neurotransmitters and boutons and all the rest of it. So, why would anyone suppose that between these two levels there is also a level of mental processes which are computational processes? And indeed why would anyone suppose that it's at that level that the brain performs those functions that we regard as essential to the survival of the organism – namely the functions of information processing?

Well, there are several reasons: First of all let me mention one which is somewhat disreputable, but I think is actually very influential. Because we do not understand the brain very well we are constantly tempted to use the latest technology as a model for trying to understand it. In my childhood we were always assured that the brain was a telephone switchboard. ('What else could it be?') I was amused to see that Sherrington, the great British neuroscientist, thought that the brain worked like a telegraph system. Freud often compared the brain to hydraulic and electro-magnetic systems. Leibniz compared it to a mill, and I am told that some of the ancient Greeks thought the brain functions like a catapult. At present, obviously, the metaphor is the digital computer.

And this, by the way, fits in with the general exaggerated guff we hear nowadays about computers and robots. We are frequently assured by the popular press that we are on the verge of having household robots that will do all of the housework, babysit our children, amuse us with lively conversation, and take care of us in our old age. This of course is so much nonsense. We are nowhere near being able to produce robots that can do any of these things. And indeed successful robots have been confined to very restricted tasks, in very limited contexts such as automobile production lines.

Well, let's get back to the serious reasons that people have for supposing that congnitivism is true. First of all, they suppose that they actually have some psychological evidence that it's true. There are two kinds of evidence. The first comes from reaction-time experiments, that is, experiments which show that different intellectual tasks take different amounts of time for people to perform. The idea here is that if the differences in the amount of time that people take are parallel to the differences in the time a computer would take, then that is at least evidence that the human system is working on the same principles as a computer. The second sort of evidence comes from linguistics, especially from the work of Chomsky and his colleagues on generative grammar. The idea here is that the formal rules of grammar which people follow when they speak a language are like the formal rules which a computer follows.

I will not say much about the reaction-time evidence, because I think everyone agrees that it is quite inconclusive and subject to a lot of different interpretations. I will say something about the linguistic evidence.

However, underlying the computational interpretation of both kinds of evidence is a much deeper, and I believe, more influential reason for accepting cognitivism. The second reason is a general thesis which the two kinds of evidence are supposed to exemplify, and it goes like this: Because we can design computers that follow rules when they process information, and because apparently human beings also follow rules when they think, then there is some unitary sense in which the brain and the computer are functioning in a similar – and indeed maybe the same – fashion.

The third assumption that lies behind the cognitivist research program is an old one. It goes back as far as Leibniz and probably as far as Plato. It is the assumption that a mental achievement must have theoretical causes. It is the assumption that if the output of a system is *meaningful*, in the sense that, for example, our ability to learn a language or our ability to recognise faces is a meaningful cognitive ability, then there must be some theory, internalised somehow in our brains, that underlies this ability.

Finally, there's another reason why people adhere to the cognitivist research program, especially if they are philosophically inclined. They can't see any other way to understand the relationship between the mind and the brain. Since we understand the relation of the computer program to the computer hardware, it provides an excellent model, maybe the only model, that will enable us to explain the relations between the mind and the brain. I have already answered this claim in the first chapter, so I don't need to discuss it further here.

Well, what shall we make of these arguments for cognitivism? I don't believe that I have a knockdown refutation of cognitivism in the way that I believe I have one of strong AI. But I do believe that if we examine the arguments that are given in favour of cognitivism, we will see that they are very weak. And indeed, an exposure of their weaknesses will enable us to understand several important differences between the way human beings behave and the way computers function.

Let's start with the notion of rule-following. We are told that human beings follow rules, and that computers follow rules. But, I want to argue that there is a crucial difference. In the case of human beings, whenever we follow a rule, we are being guided by the actual content or the meaning of the rule. In the case of human rule-following, meanings cause behaviour. Now of course, they don't cause the behaviour all by themselves, but they certainly play a causal role in the production of the behaviour. For example, consider the rule: Drive on the left-hand side of the road in Great Britain. Now, whenever I come to Britain I have to remind myself of this rule. How does it work? To say that I am obeying the rule is to say that the meaning of that rule, that is, its semantic content, plays some kind of causal role in the production of what I actually do. Notice that there are lots of other rules that would describe what's happening. But they are not the rules that I happen to be following. So, for example, assuming that I am on a two lane road and that the steering wheel is located on the right-hand side of the car, then you could say that my behaviour is in accord with the rule: Drive in such a way that the steering wheel is nearest to the centre line of the road. Now, that is in fact a correct description of my behaviour. But that's not the rule that I follow in Britain. The rule that I follow is: Drive on the left-hand side of the road.

I want this point to be completely clear so let me give you

another example. When my children went to the Oakland Driving School, they were taught a rule for parking cars. The rule was: Manoeuvre your car toward the kerb with the steering wheel in the extreme right position until your front wheels are even with the rear wheels of the car in front of you. Then, turn the steering wheel all the way to the extreme left position. Now notice that if they are following this rule, then its meaning must play a causal role in the production of their behaviour. I was interested to learn this rule because it is not a rule that I follow. In fact, I don't follow a rule at all when I park a car. I just look at the kerb and try to get as close to the kerb as I can without bashing into the cars in front of and behind me. But notice, it might turn out that my behaviour viewed from outside, viewed externally, is identical with the behaviour of the person who is following the rule. Still, it would not be true to say of me that I was following the rule. The formal properties of the behaviour are not sufficient to show that a rule is being followed. In order that the rule be followed, the meaning of the rule has to play some causal role in the behaviour.

Now, the moral of this discussion for cognitivism can be put very simply: *In the sense in which human beings follow rules* (and incidentally human beings follow rules a whole lot less than cognitivists claim they do), *in that sense computers don't follow rules at all. They only act in accord with certain formal procedures.* The program of the computer determines the various steps that the machinery will go through; it determines how one state will be transformed into a subsequent state. And we can speak *metaphorically* as if this were a matter of following rules. But in the *literal* sense in which human beings follow rules computers do not follow rules, they only act as if they were following rules. Now such metaphors are quite harmless, indeed they are both common and useful in science. We can speak metaphorically of any system as if it were following rules, the solar system for example. The metaphor only becomes harmful if it is confused with the literal sense. It is OK to use a psychological metaphor to explain the computer. The confusion comes when you take the metaphor literally and use the metaphorical computer sense of rule-following to try to explain the psychological sense of rule-following, on which the metaphor was based in the first place.

And we are now in a position to say what was wrong with the linguistic evidence for cognitivism. If it is indeed true that people follow rules of syntax when they talk, that doesn't show that they behave like digital computers because, in the sense in which they follow rules of syntax, the computer doesn't follow rules at all. It only goes through formal procedures.

So we have two senses of rule following, a literal and a metaphorical. And it is very easy to confuse the two. Now I want to apply these lessons to the notion of information-processing. I believe the notion of information-processing embodies a similar massive confusion. The idea is that since I process information when I think, and since my calculating machine processes information when it takes something as input, transforms it, and produces information as output, then there must be some unitary sense in which we are both processing information. But that seems to me obviously false. The sense in which I do information-processing when I think is the sense in which I am consciously or unconsciously engaged in certain mental processes. But in that sense of information-processing, the calculator does not do information-processing, since it does not have any mental processes at all. It simply mimics, or simulates the formal features of mental processes that I have. That is, even if the steps that the calculator goes through are formally the same as the steps that I go through, it would not show that the machine does anything at all like what I do, for the very simple reason that the calculator has no mental phenomena. In adding 6 and 3, it doesn't know that the numeral '6' stands for the number six, and that the numeral '3' stands for the number three, and that the plus sign stands for the operation of addition. And that's for the very simple reason that it doesn't know anything. Indeed, that is why we have calculators. They can do calculations faster and more accurately than we can without having to go through any mental effort to do it. In the sense in which we have to go through information-processing, they don't.

We need, then, to make a distinction between two senses of the notion of information-processing. Or at least, two radically different kinds of information-processing. The first kind, which I will call 'psychological information-processing' involves mental states. To put it at its crudest: When people perform mental operations, they actually think, and thinking characteristically involves processing information of one kind or another. But there is another sense of information-processing in which there are no mental states at all. In these cases, there are processes which are *as if* there were some mental information-processing going on. Let us call these second kinds of cases of information-processing 'as if' forms of information-processing. It is perfectly harmless to use both of these two kinds of mental ascriptions provided we do not confuse them. However, what we find in cognitivism is a persistent confusion of the two.

Now once we see this distinction clearly, we can see one of the most profound weaknesses in the cognitivist argument. From the fact that I do information-processing when I think, and the fact that the computer does information-processing – even information-processing which may simulate the formal features of my thinking – it simply doesn't follow that there is anything psychologically relevant about the computer program. In order to show psychological relevance, there would have to be some independent argument that the 'as if' computational information-processing is psychologically relevant. The notion of information-processing is being used to mask this confusion, because one expression is being used to cover two quite distinct phenomena. In short, the confusion that we found about rule-following has an exact parallel in the notion of information-processing.

However, there is a deeper and more subtle confusion involved in the notion of information-processing. Notice that in the 'as if' sense of information-processing, any system whatever can be described as if it were doing information-processing, and indeed, we might even use it for gathering information. So, it isn't just a matter of using calculators and computers. Consider, for example, water running downhill. Now, we can describe the water as if it were doing information-processing. And we might even use it to get information. We might use it, for example, to get information about the line of least resistance in the contours of the hill. But it doesn't follow from that that there is anything of psychological relevance about water running downhill. There's no psychology at all to the action of gravity on water.

But we can apply the lessons of this point to the study of the brain. It's an obvious fact that the brain has a level of real psychological information processes. To repeat, people actually think, and thinking goes on in their brains. Furthermore, there are all sorts of things going on in the brain at the neurophysiological level that actually cause our thought processes. But many people suppose that in addition to these two levels, the level of naive psychology and the level of neurophysiology, there must be some additional level of computational information-processing. Now why do they suppose that? I believe that it is partly because they confuse the psychologically real level of information-processing with the possibility of giving 'as if' information-processing descriptions of the processes going on in the brain. If you talk about water running downhill, everyone can see that it is psychologically irrelevant. But it is harder to see that exactly the same point applies to the brain.

What is psychologically relevant about the brain is the facts that it contains psychological processes and that it has a neurophysiology that causes and realises these processes. But the fact that we can describe other processes in the brain from an 'as if' information-processing point of view, by itself provides no evidence that these are psychologically real or even psychologically relevant. Once we are talking about the inside of the brain, it's harder to see the confusion, but it's exactly the same confusion as the confusion of supposing that because water running downhill does 'as if' information-processing, there is some hidden psychology in water running downhill.

The next assumption to examine is the idea that behind all meaningful behaviour there must be some internal theory. One finds this assumption in many areas and not just in cognitive psychology. So for example, Chomsky's search for a universal grammar is based on the assumption that if there are certain features common to all languages and if these features are constrained by common features of the human brain, then there must be an entire complex set of rules of universal grammar in the brain. But a much simpler hypothesis would be that the physiological structure of the brain constrains possible grammars without the intervention of an intermediate level of rules or theories. Not only is this hypothesis simpler, but also the very existence of universal features of language constrained by innate features of the brain suggests that the neurophysiological level of description is enough. You don't need to suppose that there are any rules on top of the neurophysiological structures.

A couple of analogies, I hope, will make this clear. It is a simple fact about human vision that we can't see infra-red or ultra-violet. Now is that because we have a universal rule of visual grammar that says: 'Don't see infra-red or ultra-violet.'? No, it is obviously because our visual apparatus simply is not sensitive to these two ends of the spectrum. Of course we could describe ourselves *as if* we were following a rule of visual grammar, but all the same, we are not. Or, to take another example, if we tried to do a theoretical analysis of the human ability to stay in balance while walking, it might look as if there were some more or less complex mental processes going on, as if taking in cues of various kinds we solved a series of quadratic equations, unconsciously of course, and these enabled us to walk without falling over. But we actually know that this sort of mental theory is not necessary to account for the achievement of walking without falling over. In fact, it is done in a very large part by fluids in the inner ear that simply do no calculating at all. If you spin around enough so as to upset the fluids, you are likely to fall over. Now I want to suggest that a great deal of our cognitive achievements may well be like that. The brain just does them. We have no good reason for supposing that in addition to the level of our mental states and the level of our neurophysiology there is some unconscious calculating going on.

Consider face recognition. We all recognise the faces of our friends, relatives and acquaintances quite effortlessly; and indeed we now have evidence that certain portions of the brain are specialised for face recognition. How does it work? Well, suppose we were going to design a computer that could recognise faces as we do. It would carry out quite a computational task, involving a lot of calculating of geometrical and topological features. But is that any evidence that the way we do it involves calculating and computing? Notice that when we step in wet sand and make a footprint, neither our feet nor the sand does any computing. But if we were going to design a program that would calculate the topology of a footprint from information about differential pressures on the sand, it would be a fairly complex computational task. The fact that a computational simulation of a natural phenomenon involves complex information-processing does not show that the phenomenon itself involves such processing. And it may be that facial recognition is as simple and as automatic as making footprints in the sand.

Indeed, if we pursue the computer analogy consistently, we find that there are a great many things going on in the computer that are not computational processes either. For example, in the case of some calculators, if you ask: 'How does the calculator multiply seven times three?', the answer is: 'It adds three to itself seven times.' But if you then ask: 'And how does it add three to itself?', there isn't any computational answer to that; it is just done in the hardware. So the answer to the question is: 'It just does it.' And I want to suggest that for a great many absolutely fundamental abilities, such as our ability to see or our ability to learn a language, there may not be any theoretical mental level underlying those abilities: the brain just does them. We are neurophysiologically so constructed that the assault of photons on our photoreceptor cells enables us to see, and we are neurophysiologically so constructed that the stimulation of hearing other people talk and interacting with them will enable us to learn a language.

Now I am not saying that rules play no role in our behaviour. On the contrary, rules of language or rules of games, for example, seem to play a crucial role in the relevant behaviour. But I am saying that it is a tricky question to decide which parts of behaviour are rule-governed and which are not. And we can't just assume all meaningful behaviour has some system of rules underlying it.

Perhaps this is a good place to say that though I am not optimistic about the overall research project of cognitivism, I do think that a lot of insights are likely to be gained from the effort, and I certainly do not want to discourage anyone from trying to prove me wrong. And even if I am right, a great deal of insight can be gained from failed research projects; behaviourism and Freudian psychology are two cases in point. In the case of cognitivism, I have been especially impressed by David Marr's work on vision and by the work of various people on 'natural language understanding', that is, on the effort to get computers to simulate the production and interpretation of ordinary human speech.

I want to conclude this chapter on a more positive note by saying what the implications of this approach are for the study of the mind. As a way of countering the cognitivist picture, let me present an alternative approach to the solution of the problems besetting the social sciences. Let's abandon the idea that there is a computer program between the mind and the brain. Think of the mind and mental processes as biological phenomena which are as biologically based as growth or digestion or the secretion of bile. Think of our visual experience, for example, as the end product of a series of events that begins with the assault of photons on the retina and ends somewhere in the brain. Now there will be two gross levels of description in the causal account of how vision takes place in animals. There will be first a level of the neurophysiology; a level at which we can discuss individual neurons, synapses, and action potentials. But within this neurophysiological level there will be lower and higher levels of description. It is not necessary to confine ourselves solely to neurons and synapses. We can talk about the behaviour of groups or modules of neurons, such as the different levels of types of neurons in the retina or the columns in the cortex; and we can talk about the performance of the neurophysiological systems at much greater levels of complexity; such as the role of the striate cortex in vision, or the role of zones 18 and 19 in the visual cortex, or the relationship between the visual cortex and the rest of the brain in processing visual stimuli. So within the neurophysiological level there will be a series of levels of description, all of them equally neurophysiological.

Now in addition to that, there will also be a mental level of description. We know, for example, that perception is a function of expectation. If you expect to see something, you will see it much more readily. We know furthermore that perception can be affected by various mental phenomena. We know that mood and emotion can affect how and what one perceives. And again, within this mental level, there will be different levels of description. We can talk not only about how perception is affected by individual beliefs and desires, but also about how it is affected by such global mental phenomena as the person's background abilities, or his general world outlook. But in addition to the level of the neurophysiology, and the level of intentionality, we don't need to suppose there is another level; a level of digital computational processes.

And there is no harm at all in thinking of both the level of mental states and the level of neurophysiology as information-processing, provided we do not make the confusion of supposing that the real psychological form of information-processing is the same as the 'as if'.

To conclude then, where are we in our assessment of the cognitivist research program? Well I have certainly not demonstrated that it is false. It might turn out to be true. I think its chances of success are about as great as the chances of success of behaviourism fifty years ago. That is to say, I think its chances of success are virtually nil. What I have done to argue for this, however, is simply the following three things: first, I have suggested that once you have laid bare the basic assumptions behind cognitivism, their implausibility is quite apparent. But these assumptions are, in large part, very deeply seated in our intellectual culture, some of them are very hard to root out or even to become fully conscious of. My first claim is that once we fully understand the nature of the assumptions, their implausibility is manifest. The second point I have made is that we do not actually have sufficient empirical evidence for supposing that these assumptions are true. Since the interpretation of the existing evidence rests on an ambiguity in certain crucial motions such as those of information processing and rule following. And third, I have presented an alternative view, both in this chapter and the first chapter, of the relationship between the brain and the mind; a view that does not require us to postulate any intermediate level of algorithmic computational processes mediating between the neurophysiology of the brain and the intentionality of the mind. The feature of that picture which is important for this discussion is that in addition to a level of mental states, such as beliefs and desires, and a level of neurophysiology, there is no other level, no gap filler is needed between the mind and the brain, because there is no gap to fill. The computer is probably no better and no worse as a metaphor for the brain than earlier mechanical metaphors. We learn as much about the brain by saying it's a computer as we do by saying it's a telephone switchboard, a telegraph system, a water pump, or a steam engine.

Suppose no one knew how clocks worked. Suppose it was frightfully difficult to figure out how they worked, because, though there were plenty around, no one knew how to build one, and efforts to figure out how they worked tended to destroy the clock. Now suppose a group of researchers said, 'We will understand how clocks work if we design a machine that is functionally the equivalent of a clock, that keeps time just as well as a clock.' So they designed an hour glass and claimed: 'Now we understand how clocks work,' or perhaps: 'If only we could get the hour glass to be just as accurate as a clock we would at last understand how clocks work.' Substitute 'brain' for 'clock' in this parable, and substitute 'digital computer program' for 'hour glass' and the notion of intelligence for the notion of keeping time and you have the contemporary situation in much (not all!) of artificial intelligence and cognitive science.

My overall objective in this investigation is to try to answer some of the most puzzling questions about how human beings fit into the rest of the universe. In the first chapter I tried to solve the 'mind-body problem'. In the second I disposed of some extreme claims that identify human beings with digital computers. In this one I have raised some doubts about the cognitivist research program. In the second half of the book, I want to turn my attention to explaining the structure of human actions, the nature of the social sciences, and the problems of the freedom of the will.

# FOUR
## THE STRUCTURE OF ACTION

The purpose of this chapter is to explain the structure of human action. I need to do that for several reasons. I need to show how the nature of action is consistent with my account of the mind-body problem and my rejection of artificial intelligence, contained in earlier chapters. I need to explain the mental component of action and show how it relates to the physical component. I need to show how the structure of action relates to the explanation of action. And I need to lay a foundation for the discussion of the nature of the social sciences and the possibility of the freedom of the will, which I will discuss in the last two chapters.

If we think about human actions, we immediately find some striking differences between them and other events in the natural world. At first, it is tempting to think that types of actions or behaviour can be identified with types of bodily movements. But that is obviously wrong. For example, one and the same set of human bodily movements might constitute a dance, or signalling, or exercising, or testing one's muscles, or none of the above. Furthermore, just as one and the same set of types of physical movements can constitute completely different kinds of actions, so one type of action can be performed by a vastly different number of types of physical movements. Think, for example, of sending a message to a friend. You could write it out on a sheet of paper. You could type it. You could send it by messenger or by telegram. Or you could speak it over the telephone. And indeed, each of these ways of sending the same message could be accomplished in a variety of physical movements. You could write the note with your left hand or your right hand, with your toes, or even by holding the pencil between your teeth. Furthermore, another odd feature about actions which makes them differ from events generally is that actions seem to have preferred descriptions. If I am going for a walk to Hyde Park, there are any number of other things that are happening in the course of my walk, but their descriptions do not describe my intentional actions, because in acting, what I am doing depends in large part on what I think I am doing. So for example, I am also moving in the general direction of Patagonia, shaking the hair on my head up and down, wearing out my shoes, and moving a lot of air molecules. However, none of these other descriptions seems to get at what is essential about this action, as the action it is.

A third related feature of actions is that a person is in a special position to know what he is doing. He doesn't have to observe himself or conduct an investigation to see which action he is performing, or at least is trying to perform. So, if you say to me: 'Are you trying to walk to Hyde Park or trying to get closer to Patagonia?' I have no hesitation in giving an answer even though the physical movements that I make might be appropriate for either answer.

It is also a remarkable fact about human beings that quite effortlessly we are able to identify and explain the behaviour of ourselves and of other people. I believe that this ability rests on our unconscious mastery of a certain set of principles, just as our ability to recognise something as a sentence of English rests on our having an unconscious mastery of the principles of English grammar. I believe there is a set of principles that we presuppose when we say such ordinary commonsense things as that, for example, Basil voted for the Tories because he thought that they would cure the problem of inflation, or Sally moved from Birmingham to London because she thought the job opportunities were better there, or even such simple things as that the man over there making such strange movements is in fact sharpening an axe, or polishing his shoes.

It is common for people who recognise the existence of these theoretical principles to sneer at them by saying that they are just a folk theory and that they should be supplanted by some more scientific account of human behaviour. I am suspicious of this claim just as I would be suspicious of a claim that said we should supplant our implicit theory of English grammar, the one we acquire by learning the language. The reason for my suspicion in each case is the same: using the implicit theory is part of performing the action in the same way that using the rules of grammar is part of speaking. So though we might add to it or discover all sorts of interesting additional things about language or about behaviour, it is very unlikely that we could replace that theory which is implicit and partly constitutive of the phenomenon by some external 'scientific' account of that very phenomenon.

Aristotle and Descartes would have been completely familiar with most of our explanations of human behaviour, but not with our explanations of biological and physical phenomena. The reason usually adduced for this is that Aristotle and Descartes had both a primitive theory of biology and physics on the one hand, and a primitive theory of human behaviour on the other; and that while we have advanced in biology and physics, we have made no comparable advance in the explanation of human behaviour. I want to suggest an alternative view. I think that Aristotle and Descartes, like ourselves, already had a sophisticated and complex theory of human behaviour. I also think that many supposedly scientific accounts of human behaviour, such as Freud's, in fact employ rather than replace the principles of our implicit theory of human behaviour.

To summarise what I have said so far: There is more to types of action than types of physical movements, actions have preferred descriptions, people know what they are doing without observation, and the principles by which we identify and explain action are themselves part of the actions, that is, they are partly *constitutive* of actions. I now want to give a brief account of what we might call the structure of behaviour.

In order to explain the structure of human behaviour, I need to introduce one or two technical terms. The key notion in the structure of behaviour is the notion of intentionality. To say that a mental state has intentionality simply means that it is about something. For example, a belief is always a belief that such and such is the case, or a desire is always a desire that such and such should happen or be the case. Intending, in the ordinary sense, has no special role in the theory of intentionality. Intending to do something is just one kind of intentionality along with believing, desiring, hoping, fearing and so on.

An intentional state like a belief, or a desire, or an intention in the ordinary sense, characteristically has two components. It has what we might call its 'content', which makes it about something, and its 'psychological mode' or 'type'. The reason we need this distinction is that you can have the same content in different types. So, for example, I can want to leave the room, I can believe that I will leave the room, and I can intend to leave the room. In each case, we have the same content, that I will leave the room; but in different psychological modes or types: belief, desire, and intending respectively.

Furthermore, the content and the type of the state will serve to relate the mental state to the world. That after all is why we have minds with mental states: to represent the world to ourselves; to represent how it is, how we would like it to be, how we fear it may turn out, what we intend to do about it and so on. This has the consequence that our beliefs will be true if they match the way the world is, false if they don't; our desires will be fulfilled or frustrated, our intentions carried out or not carried out. In general, then, intentional states have 'conditions of satisfaction'. Each state itself determines under what conditions it is true (if, say, it is a belief) or under what conditions it is fulfilled (if, say, it is a desire) and under what conditions it is carried out (if it is an intention). In each case the mental state represents its own conditions of satisfaction.

A third feature to notice about such states is that sometimes they cause things to happen. For example, if I want to go to the movies, and I do go to the movies, normally my desire will cause the very event that it represents, my going to the movies. In such cases there is an internal connection between the cause and the effect, because the cause is a representation of the very state of affairs that it causes. The cause both represents and brings about the effect. I call such kinds of cause and effect relations, cases of 'intentional causation'. Intentional causation as we will see, will prove crucial both to the structure and to the explanation of human action. It is in various ways quite different from the standard textbook accounts of causation, where for example one billiard ball hits another billiard ball, and causes it to move. For our purposes the essential thing about intentional causation is that in the cases we will be considering the mind brings about the very state of affairs that it has been thinking about.

To summarise this discussion of intentionality, there are three features that we need to keep in mind in our analysis of human behaviour: First, intentional states consist of a content in a certain mental type. Second, they determine their conditions of satisfaction, that is, they will be satisfied or not depending on whether the world matches the content of the state. And third, sometimes they cause things to happen, by way of intentional causation to bring about a match – that is, to bring about the state of affairs they represent, their own conditions of satisfaction.

Using these ideas I'll now turn to the main task of this chapter. I promised to give a very brief account of what might be called the structure of action, or the structure of behaviour. By behaviour here, I mean voluntary, intentional human behaviour. I mean such things as walking, running, eating, making love, voting in elections, getting married, buying and selling, going on a vacation, working on a job. I do not mean such things as digesting, growing older, or snoring. But even restricting ourselves to intentional behaviour, human activities present us with a bewildering variety of types. We will need to distinguish between individual behaviour and social behaviour; between collective social behaviour and individual behaviour within a social collective; between doing something for the sake of something else, and doing something for its own sake. Perhaps most difficult of all, we need to account for the melodic sequences of behaviour through the passage of time. Human activities, after all, are not like a series of still snapshots, but something more like the movie of our life.

I can't hope to answer all of these questions, but I do hope in the end that what I say will seem like a commonsense account of the structure of action. If I am right, what I say should seem obviously right. But historically what I think of as the commonsense account has not seemed obvious. For one thing, the behaviourist tradition in philosophy and psychology has led many people to neglect the mental component of actions. Behaviourists wanted to define actions, and indeed all of our mental life, in terms of sheer physical movements. Somebody once characterised the behaviourist approach, justifiably in my view, as feigning anaesthesia. The opposite extreme in philosophy has been to say that the only acts we ever perform are inner mental acts of volition. On this view, we don't strictly speaking ever raise our arms. All we do is 'volit' that our arms go up. If they do go up, that is so much good luck, but not really our action.

Another problem is that until recently the philosophy of action was a somewhat neglected subject. The Western tradition has persistently emphasised knowing as more important than doing. The theory of knowledge and meaning has been more central to its concerns than the theory of action. I want now to try to draw together both the mental and the physical aspects of action.

An account of the structure of behaviour can be best given by stating a set of principles. These principles should explain both the mental and physical aspects of action. In presenting them, I won't be discussing where our beliefs, desires, and so on come from. But I will be explaining how they figure in our behaviour.

I think the simplest way to convey these principles is just to state them and then try to defend them. So here goes.

*Principle 1: Actions characteristically consist of two components, a mental component and a physical component.*

Think, for example, of pushing a car. On the one hand, there are certain conscious experiences of effort when you push. If you are successful, those experiences will result in the movement of your body and the corresponding movement of the car. If you are unsuccessful, you will still have had at least the mental component, that is, you will still have had an experience of trying to move the car with at least some of the physical components. There will have been muscle tightenings, the feeling of pressure against the car, and so on. This leads to

*Principle 2: The mental component is an intention.* It has intentionality – it is about something. It determines what counts as success or failure in the action; and if successful, it causes the bodily movement which in turn causes the other movements, such as the movement of the car, which constitute the rest of the action. In terms of the theory of intentionality that we just sketched, the action consists of two components, a mental component and a physical component. If successful, the mental component causes the physical component and it represents the physical component. This form of causation I call 'intentional causation'.

The best way to see the nature of the different components of an action is to carve each component off and examine it separately. And in fact, in a laboratory, it's easy enough to do that. We already have in neurophysiology experiments, done by Wilder Penfield of Montreal, where by electrically stimulating a certain portion of the patient's motor cortex, Penfield could cause the movement of the patient's limbs. Now, the patients were invariably surprised at this, and they characteristically said such things as: 'I didn't do that – you did it.' In such a case, we have carved off the bodily movement without the intention. Notice that in such cases the bodily movements might be the same as they are in an intentional action, but it seems quite clear that there is a difference. What's the difference? Well, we also have experiments going back as far as William James, where we can carve off the mental component without the corresponding physical component of the action. In the James case, a patient's arm is anaesthetised, and it is held at his side in a dark room, and he is then ordered to raise it. He does what he thinks is obeying the order, but is later quite surprised to discover that his arm didn't go up. Now in that case, we carve off the mental component, that is to say the intention, from the bodily movement. For the man really did have the intention. That is, we can truly say of him, he genuinely did try to move his arm.

Normally these two components come together. We usually have both the intention and the bodily movement, but they are not independent. What our first two principles try to articulate is how they are related. The mental component as part of its conditions of satisfaction has to both *represent* and *cause* the physical component. Notice, incidentally, that we have a fairly extensive vocabulary, of 'trying', and 'succeeding', and 'failing', of 'intentional' and 'unintentional', of 'action' and 'movement', for describing the workings of these principles.

*Principle 3: The kind of causation which is essential to both the structure of action and the explanation of action is intentional causation.* The bodily movements in our actions are caused by our intentions. Intentions are causal because they make things happen; but they also have contents and so can figure in the process of logical reasoning. They can be both causal and have logical features because the kind of causation we are talking about is mental causation or intentional causation. And in intentional causation mental contents affect the world. The whole apparatus works because it is realised in the brain, in the way I explained in the first chapter.

The form of causation that we are discussing here is quite different from the standard form of causation as described in philosophical textbooks. It is not a matter of regularities or covering laws or constant conjunctions. In fact, I think it's much closer to our commonsense notion of causation, where we just mean that something makes something else happen. What is special about intentional causation is that it is a case of a mental state making something else happen, and that something else is the very state of affairs represented by the mental state that causes it.

*Principle 4: In the theory of action, there is a fundamental distinction between those actions which are premeditated, which are a result of some kind of planning in advance, and those actions which are spontaneous, where we do something without any prior reflection.* And corresponding to this distinction, we need a distinction between *prior intentions*, that is, intentions formed before the performance of an action, and *intentions in action*, which are the intentions we have while we are actually performing an action.

A common mistake in the theory of action is to suppose that all intentional actions are the result of some sort of deliberation, that they are the product of a chain of practical reasoning. But obviously, many things we do are not like that. We simply do something without any prior reflection. For example, in a normal conversation, one doesn't reflect on what one is going to say next, one just says it. In such cases, there is indeed an intention, but it is not an intention formed prior to the performance of the action. It is what I call an intention in action. In other cases, however, we do form prior intentions. We reflect on what we want and what is the best way to achieve it. This process of reflection (Aristotle called it 'practical reasoning'), characteristically results either in the formation of a prior intention, or, as Aristotle also pointed out, sometimes it results in the action itself.

*Principle 5: The formation of prior intentions is, at least generally, the result of practical reasoning. Practical reasoning is always reasoning about how best to decide between conflicting desires.* The motive force behind most human (and animal) action is desire. Beliefs function only to enable us to figure out how best to satisfy our desires. So, for example, I want to go to Paris, and I believe that the best way, all things considered, is to go by plane, so I form the intention to go by plane. That's a typical and commonsense piece of practical reasoning. But practical reasoning differs crucially from theoretical reasoning, from reasoning about what is the case, in that practical reasoning is always about how best to decide among the various conflicting desires we have. So, for example, suppose I do want to go to Paris, and I figure that the best way to go is to go by plane. Nonetheless, there is no way I can do that without frustrating a large number of other desires I have. I don't want to spend money; I don't want to stand in queues at airports; I don't want to sit in airplane seats; I don't want to eat airplane food; I don't want people to put their elbow where I'm trying to put my elbow; and so on indefinitely. Nonetheless, in spite of all of the desires that will be frustrated if I go to Paris by plane, I may still reason that it's best, all things considered, to go to Paris by plane. This is not only typical of practical reasoning, but I think it's universal in practical reasoning that practical reasoning concerns the adjudication of conflicting desires.

The picture that emerges from these five principles, then, is that the mental energy that powers action is an energy that works by intentional causation. It is a form of energy whereby the cause, either in the form of desires or intentions, represents the very state of affairs that it causes.

Now let's go back to some of those points about action that we noticed at the beginning, because I think we have assembled enough pieces to explain them. We noticed that actions had preferred descriptions, and that, in fact, common sense enabled us to identify what the preferred descriptions of actions were. Now we can see that the preferred description of an action is determined by the intention in action. What the person is really doing, or at least what he is trying to do, is entirely a matter of what the intention is that he is acting with. For example, I know that I am trying to get to Hyde Park and not trying to get closer to Patagonia, because that's the intention with which I am going for a walk. And I know this without *observation* because the knowledge in question is not knowledge of my external behaviour, but of my inner mental states.

This furthermore explains some of the logical features about the explanations that we give of human action. To explain an action is to give its causes. Its causes are psychological states. Those states relate to the action either by being steps in the practical reasoning that led to the intentions or the intentions themselves. The most important feature of the explanation of action, however, is worth the statement as a separate principle, so let's call it

*Principle 6: The explanation of an action must have the same content as was in the person's head when he performed the action or when he reasoned toward his intention to perform the action. If the explanation is really explanatory, the content that causes behaviour by way of intentional causation must be identical with the content in the explanation of the behaviour.*

In this respect actions differ from other natural events in the world, and correspondingly, their explanations differ. When we explain an earthquake or a hurricane, the content in the explanation only has to represent what happened and why it happened. It doesn't actually have to cause the event itself. But in explaining human behaviour, the cause and the explanation both have contents and the explanation only explains because it has the same content as the cause.

So far we have been talking as if people just had intentions out of the blue. But, of course, that is very unrealistic. And we now need to introduce some complexities which will get our analysis at least a bit closer to the affairs of real life. No one ever just has an intention, just like that by itself. For example, I have an intention to drive to Oxford from London: I may have that quite spontaneously, but nonetheless I must still have a series of other intentional states. I must have a belief that I have a car and a belief that Oxford is within driving distance. Furthermore, I will characteristically have a desire that the roads won't be too crowded and a wish that the weather won't be too bad for driving. Also (and here it gets a little closer to the notion of the explanation of action) I will characteristically not just drive to Oxford, but drive to Oxford for some purpose. And if so, I will characteristically engage in practical reasoning – that form of reasoning that leads not to beliefs or conclusions of arguments, but to intentions and to actual behaviour. And when we understand this form of reasoning, we will have made a great step toward understanding the explanation of actions. Let us call the other intentional states that give my intentional state the particular meaning that it has, let us call all of them 'the network of intentionality'. And we can say by way of a general conclusion – let's call this

*Principle 7: Any intentional state only functions as part of a network of other intentional states. And by 'functions' here, I mean that it only determines its conditions of satisfaction relative to a whole lot of other intentional states.*

Now, when we begin to probe the details of the network, we discover another interesting phenomenon. And that is simply that the activities of our mind cannot consist in mental states, so to speak, right down to the ground. Rather, our mental states only function in the way they do because they function against a background of capacities, abilities, skills, habits, ways of doing things, and general stances toward the world that do not themselves consist in intentional states. In order for me so much as to form the intention to drive to Oxford, I have to have the ability to drive. But the ability to drive doesn't itself consist in a whole lot of other intentional states. It takes more than a bunch of beliefs and desires in order to be able to drive. I actually have to have the skill to do it. This is a case where my knowing how is not just a matter of knowing that. Let us call the set of skills, habits, abilities, etc. against which intentional states function 'the background of intentionality'. And to the thesis of the network, namely that any intentional state only functions as a part of a network, we will add the thesis of the background – call it

*Principle 8: The whole network of intentionality only functions against a background of human capacities that are not themselves mental states.*

I said that many supposedly scientific accounts of behaviour try to escape from or surpass this commonsense model that I have been sketching. But in the end there's no way I think they can do that, because these principles don't just describe the phenomena: they themselves partly go to make up the phenomena. Consider, for example, Freudian explanations. When Freud is doing his metapsychology, that is, when he is giving the theory of what he is doing, he often uses scientific comparisons. There are a lot of analogies between psychology and electromagnetism or hydraulics, and we are to think of the mind as functioning on the analogy of hydraulic principles, and so on. But when he is actually examining a patient, and he is actually describing the nature of some patient's neurosis, it is surprising how much the explanations he gives are commonsense explanations. Dora behaves the way she does because she is in love with Herr K, or because she's imitating her cousin who has gone off to Mariazell. What Freud adds to common sense is the observation that often the mental states that cause our behaviour are unconscious. Indeed, they are repressed. We are often resistant to admitting to having certain intentional states because we are ashamed of them, or for some other reason. And secondly, he also adds a theory of the transformations of mental states, how one sort of intentional state can be transformed into another. But with the addition of these and other such accretions, the Freudian form of explanation is the same as the commonsense forms. I suggest that common sense is likely to persist even as we acquire other more scientific accounts of behaviour. Since the structure of the explanation has to match the structure of the phenomena explained, improvements in explanation are not likely to have new and unheard-of structures.

In this chapter I have tried to explain how and in what sense behaviour both contains and is caused by internal mental states. It may seem surprising that much of psychology and cognitive science have tried to deny these relations. In the next chapter, I am going to explore some of the consequences of my view of human behaviour for the social sciences. Why is it that the social sciences have suffered failures and achieved the successes that they have, and what can we reasonably expect to learn from them?

# FIVE

# PROSPECTS FOR THE SOCIAL SCIENCES

In this Chapter I want to discuss one of the most vexing intellectual problems of the present era: Why have the methods of the natural sciences not given us the kind of payoff in the study of human behaviour that they have in physics and chemistry? And what sort of 'social' or 'behavioural' sciences can we reasonably expect anyhow? I am going to suggest that there are certain radical differences between human behaviour and the phenomena studied in the natural sciences. I will argue that these differences account both for the failures and the successes that we have had in the human sciences.

At the beginning I want to call your attention to an important difference between the form of commonsense explanations of human behaviour and the standard form of scientific explanation. According to the standard theory of scientific explanation, explaining a phenomenon consists in showing how its occurrence follows from certain scientific laws. These laws are universal generalisations about how things happen. For example, if you are given a statement of the relevant laws describing the behaviour of a falling body, and you know where it started from, you can actually deduce what will happen to it. Similarly if you want to explain a law, you can deduce the law from some higher level law. On this account explanation and prediction are perfectly symmetrical. You predict by deducing what will happen; you explain by deducing what has happened. Now, whatever merit this type of explanation may have in the natural sciences, one of the things I want to emphasise in this chapter is that it is quite worthless to us in explaining human behaviour. And this is not because we lack laws for explaining individual examples of human behaviour. It's because even if we had such laws, they would still be useless to us. I think I can easily get you to see that by asking you to imagine what it would be like if we actually had a 'law', that is, a universal generalisation, concerning some aspect of your behaviour.

Suppose that in the last election, you voted for the Tories, and suppose that you voted for the Tories because you thought they would do more to solve the problem of inflation than any of the other parties. Now, suppose that that is just a plain fact about why you voted for the Tories, as it is an equally plain fact that you did vote for the Tories. Suppose furthermore that some political sociologists come up with an absolutely exceptionless universal generalisation about people who exactly fit your description – your socio-economic status, your income level, your education, your other interests, and so on. Suppose the absolutely exceptionless generalisation is to the effect that people like you invariably vote for the Tories. Now I want to ask: which explains why you voted for the Tories? Is it the reason that you sincerely accept? Or the universal generalisation? I want to argue that we would never accept the generalisation as the explanation of our own behaviour. The generalisation states a regularity. Knowledge of such a regularity may be useful for prediction, but it does not explain anything about individual cases of human behaviour. Indeed it invites further explanation. For instance, why do all these people in that group vote for the Tories? An answer suggests itself. You voted for the Tories because you were worried about inflation – perhaps people in your group are particularly affected by inflation and that is why they all vote the same way.

In short, we do not accept a generalisation as explaining our own or anybody else's behaviour. If a generalisation were found, it itself would require explanation of the sort we were after in the first place. And where human behaviour is concerned, the sort of explanation we normally seek is one that specifies the mental states – beliefs, fears, hopes, desires, and so on – that function causally in the production of the behaviour, in the way that I described in the previous chapter.

Let's return to our original question: Why do we not seem to have laws of the social sciences in the sense that we have laws of the natural sciences? There are several standard answers to that question. Some philosophers point out that we don't have a science of behaviour for the same reason we don't have a science of furniture. We couldn't have such a science because there aren't any physical features that chairs, tables, desks, and all other items of furniture have in common that would enable them to fall under a common set of laws of furniture. And besides we don't really need such a science because anything we want to explain – for example, why are wooden tables solid or why does iron lawn furniture rust? – can already be explained by existing sciences. Similarly, there aren't any features that all human behaviours have in common. And besides, particular things we wish to explain can be explained by physics, and physiology, and all the rest of the existing sciences.

In a related argument some philosophers point out that perhaps our concepts for describing ourselves and other human beings don't match the concepts of such basic sciences as physics and chemistry in the right way. Perhaps, they suggest, human science is like a science of the weather. We have a science of the weather, meteorology, but it is not a strict science because the things that interest us about the weather don't match the *natural* categories we have for physics. Such weather concepts as 'bright spots over the Midlands' or 'partly cloudy in London' are not systematically related to the concepts of physics. A powerful expression of this sort of view is in Jerry Fodor's work. He suggests that special sciences like geology or meteorology are about features of the world that can be realised in physics in a variety of ways and that this loose connection between the special science and the more basic science of physics is also characteristic of the social sciences. Just as mountains and storms can be realised in different sorts of microphysical structures, so money for example can be physically realised as gold, silver or printed paper. And such disjunctive connections between the higher order phenomena and the lower order do indeed allow us to have rich sciences, but they do not allow for *strict* laws, because the form of the loose connections will permit of laws that have exceptions.

Another argument for the view that we cannot have strict laws connecting the mental and the physical is in Donald Davidson's claim that the concepts of rationality, consistency and coherence are partly constitutive of our notion of mental phenomena; and these notions don't relate systematically to the notions of physics. As Davidson says they have no 'echo' in physics. A difficulty with this view, however, is that there are lots of sciences which contain constitutive notions that similarly have no echo in physics but are nonetheless pretty solid sciences. Biology, for example, requires the concept of organism, and 'organism' has no echo in physics, but biology does not thereby cease to be a hard science.

Another view, widely held, is that the complex interrelations of our mental states prevent us from ever getting a systematic set of law connecting them to neurophysiological states. According to this view, mental states come in complex, interrelated networks, and so cannot be systematically mapped onto types of brain states. But once again, this argument is inconclusive. Suppose, for example, that Noam Chomsky is right in thinking that each of us has a complex set of rules of universal grammar programmed into our brains at birth. There is nothing about the complexity or interdependence of the rules of the universal grammar that prevents them from being systematically realised in the neurophysiology of the brain. Interdependence and complexity by themselves are not a sufficient argument against the possibility of strict psychophysical laws.

I find all of these accounts suggestive but I do not believe that they adequately capture the really radical differences between the mental and the physical sciences. The relation between sociology and economics on the one hand and physics on the other is really quite unlike the relations of for example meteorology, geology, and biology and other special natural sciences to physics; and we need to try to state exactly how. Ideally, I would like to be able to give you a step by step argument to show the limitations on the possibilities of strict social sciences, and yet show the real nature and power of these disciplines. I think we need to abandon once and for all the idea that the social sciences are like physics before Newton, and that what we are waiting for is a set of Newtonian laws of mind and society.

First, what exactly is the problem supposed to be? One might say, 'Surely social and psychological phenomena are as real as anything else. So why can't there be laws of their behaviour?' Why should there be laws of the behaviour of molecules but not laws of the behaviour of societies? Well, one way to disprove a thesis is to imagine that it is true and then show that that supposition is somehow absurd. Let's suppose that we actually had laws of society and laws of history that would enable us to predict when there would be wars and revolutions. Suppose that we could predict wars and revolutions with the same precision and accuracy that we can predict the acceleration of a falling body in a vacuum at sea level.

The real problem is this: Whatever else wars and revolutions are, they involve lots of molecule movements. But that has the consequence that any strict law about wars and revolutions would have to match perfectly with the laws about molecule movements. In order for a revolution to start on such and such a day, the relevant molecules would have to be blowing in the right direction. But if that is so, then the laws that predict the revolution will have to make the same predictions at the level of the revolutions and their participants that the laws of molecule movements make at the level of the physical particles. So now our original question can be reformulated. Why can't the laws at the higher level, the level of revolutions, match perfectly with the laws at the lower level, the level of particles? Well, to see why they can't, let's examine some cases where there really is a perfect match between the higher order laws and the lower order laws, and then we can see how these cases differ from the social cases.

One of the all-time successes in reducing the laws at one level to those of a lower level is the reduction of the gas laws – Boyle's Law and Charles's Law – to the laws of statistical mechanics. How does the reduction work? The gas laws concern the relation between pressure, temperature, and volume of gases. They predict, for example, that if you increase the temperature of a gas in a cylinder, you will increase the pressure on the walls of the cylinder. The laws of statistical mechanics concern the behaviour of masses of small particles. They predict, for example, that if you increase the rate of movement of the particles in a gas, more of the particles will hit the walls of the cylinder and will hit them harder. The reason you get a perfect match between these two sets of laws is that the explanation of temperature, pressure, and volume can be given entirely in terms of the behaviour of the particles. Increasing the temperature of the gas increases the velocity of the particles, and increasing the number and velocity of the particles hitting the cylinder increases the pressure. It follows that an increase in temperature will produce an increase in pressure. Now suppose for the sake of argument that it wasn't like that. Suppose there was no explanation of pressure and temperature in terms of the behaviour of more fundamental particles. Then any laws at the level of pressure and temperature would be miraculous. Because it would be miraculous that the way that pressure and temperature were going on coincided exactly with the way that the particles were going on, if there was no systematic relation between the behaviour of the system at the level of pressure and temperature, and the behaviour of the system at the level of the particles.

This example is a very simple case. So, let's take a slightly more complex example. It is a law of 'nutrition science' that caloric intake equals caloric output, plus or minus fat deposit. Not a very fancy law perhaps, but pretty realistic nonetheless. It has the consequence known to most of us that if you eat a lot and don't exercise enough, you get fat. Now this law, unlike the gas laws, is not grounded in any simple way in the behaviour of the particles. The grounding isn't simple – because for example there is a rather complex series of processes by which food is converted into fat deposits in live organisms. Nonetheless, there is still a grounding – however complex – of this law in terms of the behaviour of more fundamental particles. Other things being equal, when you eat a lot, the molecules will be blowing in exactly the right direction to make you fat.

We can now argue further towards the conclusion that there will be no laws of wars and revolutions in a way that there are laws of gases and of nutrition. The phenomena in the world that we pick out with concepts like war and revolution, marriage, money and property are not grounded systematically in the behaviour of elements at the more basic level in a way that the phenomena that we pick out with concepts like fat-deposit and pressure are grounded systematically in the behaviour of elements at the more basic level. Notice that it is this sort of grounding that characteristically enables us to make major advances at the higher levels of a science. The reason that the discovery of the structure of DNA is so important to biology or that the germ theory of disease is so important to medicine is that in each case it holds out the promise of systematically explaining higher-level features, such as hereditary traits and disease symptoms in terms of more fundamental elements.

But now the question arises: If the social and psychological phenomena aren't grounded in this way, why aren't they? Why couldn't they be? Granted that they are not so grounded, why not? That is, wars and revolutions, like everything else, consist of molecule movements. So why can't such social phenomena as wars and revolutions be systematically related to molecule movements in the same way that the relations between caloric inputs and fat deposits are systematic?

To see why this can't be so we have to ask what features social phenomena have that enable us to bind them into categories. What are the fundamental principles on which we categorise psychological and social phenomena? One crucial feature is this: For a large number of social and psychological phenomena the concept that names the phenomenon is itself a constituent of the phenomenon. In order for something to count as a marriage ceremony or a trade union, or property or money or even a war or revolution people involved in these activities have to have certain appropriate thoughts. In general they have to think that's what it is. So, for example, in order to get married or buy property you and other people have to think that that is what you are doing. Now this feature is crucial to social phenomena. But there is nothing like it in the biological and physical sciences. Something can be a tree or a plant, or some person can have tuberculosis even if no one thinks: 'Here's a tree, or a plant, or a case of tuberculosis', and even if no one thinks about it at all. But many of the terms that describe social phenomena have to enter into their constitution. And this has the further result that such terms have a peculiar kind of self-referentiality. 'Money' refers to whatever people use as and think of as money. 'Promise' refers to whatever people intend as and regard as promises. I am not saying that in order to have the institution of money people have to have that very word or some exact synonym in their vocabulary. Rather, they must have certain thoughts and attitudes about something in order that it counts as money and these thoughts and attitudes are part of the very definition of money.

There is another crucial consequence of this feature. The defining principle of such social phenomena set no physical limits whatever on what can count as the physical realisation of them. And this means that there can't be any systematic connections between the physical and the social or mental properties of the phenomenon. The social features in question are determined in part by the attitudes we take toward them.

The attitudes we take toward them are not constrained by the physical features of the phenomena in question. Therefore, there can't be any matching of the mental level and the level of the physics of the sort that would be necessary to make strict laws of the social sciences possible.

The main step in the argument for a radical discontinuity between the social sciences and the natural sciences depends on the mental character of social phenomena. And it is this feature which all those analogies I mentioned earlier – that is, between meteorology, biology, and geology – neglect. The radical discontinuity between the social and psychological disciplines on the one hand and the natural sciences on the other derives from the role of the mind in these disciplines.

Consider Fodor's claim that social laws will have exceptions since the phenomena at the social level map loosely or disjunctively onto the physical phenomena. Once again this does not account for the radical discontinuities I have been calling attention to. Even if this sort of disjunction had been true up to a certain point, it is always open to the next person to add to it in indefinitely many ways. Suppose money has always taken a limited range of physical forms – gold, silver, and printed paper, for example. Still, it is open to the next person or society to treat something else as money. And indeed the physical realisation doesn't matter to the properties of money as long as the physical realisation enables the stuff to be used as a medium of exchange.

'Well,' someone might object, 'in order to have rigorous social sciences we don't need a strict match between properties of things in the world. All we need is a strict match between psychological properties and features of the brain. The real grounding of economics and sociology in the physical world is not in the properties of objects we find around us, it is in the physical properties of the brain. So even if thinking that something is money is essential to its being money, still thinking that it is money may well be, and indeed on your own account is, a process in the brain. So, in order to show that there can't be any strict laws of the social sciences you have to show that there can't be any strict correlations between types of mental states and types of brain states, and you haven't shown that.'

To see why there can't be such laws, let's examine some areas where it seems likely that we will get a strict neuropsychology, strict laws correlating mental phenomena and neurophysiological phenomena. Consider pain. It seems reasonable to suppose that neurophysiological causes of pains, at least in human beings, are quite limited and specific. Indeed we discussed some of them in an earlier chapter. There seems to be no obstacle in principle to having a perfect neurophysiology of pain. But what about, say, vision? Once again it is hard to see any obstacle in principle to getting an adequate neurophysiology of vision. We might even get to the point when we could describe perfectly the neurophysiological conditions for having certain sorts of visual experiences. The experience of seeing that something is red, for instance. Nothing in my account would prevent us from having such a neurophysiological psychology.

But now here comes the hard part: though we might get systematic correlations between neurophysiology and pain or neurophysiology and the visual experience of red, we couldn't give similar accounts of the neurophysiology of *seeing* that something was money. Why not? Granted that every time you see that there is some money in front of you some neurophysiological process goes on, what is to prevent it from being the same process every time? Well, from the fact that money can have an indefinite range of physical forms it follows that it can have an indefinite range of stimulus effects on our nervous systems. But since it can have an indefinite range of stimulus patterns on our visual systems, it would once again be a miracle if they all produced exactly the same neurophysiological effect on the brain.

And what goes for *seeing* that something is money goes even more forcefully for *believing* that it is money. It would be nothing short of miraculous if every time someone believed that he was short of money, in whatever language and culture he had this belief in, it had exactly the same type of neurophysiological realisation. And that's simply because the range of possible neurophysiological stimuli that could produce that very belief is infinite. Paradoxically, the way that the mental infects the physical prevents there ever being a strict science of the mental.

Notice that, in cases when we do not have this sort of interaction between the social and the physical phenomena, this obstacle to having strict social sciences is not present. Consider the example I mentioned earlier of Chomsky's hypothesis of universal grammar. Suppose each of us has innately programmed in our brains the rules of universal grammar. Since these rules would be in the brain at birth and independent of any relations the organism has with the environment, there is nothing in my argument to prevent there being strict psycho-physical laws connecting these rules and features of the brain, however interrelated and complicated the rules might be. Again, many animals have conscious mental states, but as far as we know, they lack the self-referentiality that goes with having human languages and social institutions. Nothing in my argument would block the possibility of a science of animal behaviour. For example, there might be strict laws correlating the brain states of birds and their nest-building behaviour.

I promised to try to give you at least a sketch of a step-by-step argument. Let's see how far I got in keeping the promise. Let's set the argument out as a series of steps.

1. For there to be laws of the social sciences in the sense in which there are laws of physics there must be some systematic correlation between phenomena identified in social and psychological terms and phenomena identified in physical terms. It can be as complex as the way that weather phenomena are connected with the phenomena of physics, but there has to be some systematic correlation. In the contemporary jargon, there have to be some bridge principles between the higher and the lower levels.

2. Social phenomena are in large part defined in terms of the psychological attitudes that people take. What counts as money or a promise or a marriage is in large part a matter of what people think of as money or a promise or a marriage.

3. This has the consequence that these categories are physically open-ended. There is strictly speaking no physical limit to what we can regard as or stipulate to be money or a promise or a marriage ceremony.

4. That implies that there can't be any bridge principles between the social and the physical features of the world, that is, between phenomena described in social terms and the same phenomena described in physical terms. We can't even have the sort of loose disjunctive principles we have for weather or digestion.

5. Furthermore, it is impossible to get the right kind of bridge principles between phenomena described in mental terms and phenomena described in neurophysiological terms, that is, between the brain and the mind. And this is because there is an indefinite range of stimulus conditions for any given social concept. And this enormous range prevents concepts which aren't built into us from being realised in a way that systematically correlates mental and physical features.

I want to conclude this chapter by describing what seems to me the true character of the social sciences. The social sciences in general are about various aspects of intentionality. Economics is about the production and distribution of goods and services. Notice that the working economist can simply take intentionality for granted. He assumes that entrepreneurs are trying to make money and that consumers would prefer to be better off rather than worse off. And the 'laws' of economics then state systematic fallouts or consequences of such assumptions. Given certain assumptions, the economist can deduce that rational entrepreneurs will sell where their marginal cost equals their marginal revenue. Now notice that the law does not predict that the business man asks himself: 'Am I selling where marginal cost equals marginal revenue?' No, the law does not state the content of individual intentionality. Rather, it works out the consequences of such intentionality. The theory of the firm in microeconomics works out the consequences of certain assumptions about the desires and possibilities of consumers and businesses engaged in buying, producing and selling. Macroeconomics works out the consequences of such assumptions for whole nations and societies. But the economist does not have to worry about such questions as: 'What is money really?' or, 'What is a desire really?' If he is very sophisticated in welfare economics he may worry about the exact character of the desires of entrepreneurs and consumers, but even in such a case the systematic part of his discipline consists in working out the consequences of facts about intentionality.

Since economics is grounded not in systematic facts about physical properties such as molecular structure, in the way that chemistry is grounded in systematic facts about molecular structure, but rather in facts about human intentionality, about desires, practices, states of technology and states of knowledge, it follows that economics cannot be free of history or context. Economics as a science presupposes certain historical facts about people and societies that are not themselves part of economics. And when those facts change, economics has to change. For example, until recently the Phillips curve, a formula relating a series of factors in industrial societies, seemed to give an accurate description of economic realities in those societies. Lately it hasn't worked so well. Most economists believe that this is because it did not accurately describe reality. But they might consider: perhaps it did accurately describe reality as it was at that time. However, after the oil crises and various other events of the seventies, reality changed. Economics is a systematic formalised science, but it is not independent of context or free of history. It is grounded in human practices, but those practices are not themselves timeless, eternal or inevitable. If for some reason money had to be made of ice, then it would be a strict law of economics that money melts at temperatures above 0° centigrade. But that law would work only as long as money had to be made of ice, and besides, it doesn't tell us what is interesting to us about money.

Let us turn now to linguistics. The standard contemporary aim of linguistics is to state the various rules – phonological, syntactic, and semantic – that relate sounds and meanings in the various natural languages. An ideally complete science of linguistics would give the complete set of rules for every natural human language. I am not sure that this is the right goal for linguistics or even that it is a goal that is possible of attainment, but for the present purposes the important thing to note is that it is once again an applied science of intentionality. It is not in the least like chemistry or geology. It is concerned with specifying those historically-determined intentional contents in the minds of speakers of the various languages that are actually responsible for human linguistic competence. As with economics, the glue that binds linguistics together is human intentionality.

The upshot of this chapter can now be stated quite simply. The radical discontinuity between the social and the natural sciences doesn't come from the fact that there is only a disjunctive connection of social and physical phenomena. It doesn't even come from the fact that social disciplines have constitutive concepts which have no echo in physics nor even from the great complexity of social life. Many disciplines such as geology, biology, and meteorology have these features but that does not prevent them from being systematic natural sciences. No, the radical discontinuity derives from the intrinsically mental character of social and psychological phenomena.

The fact that the social sciences are powered by the mind is the source of their weakness vis-à-vis the natural sciences. But it is also precisely the source of their strength as social sciences.

What we want from the social sciences and what we get from the social sciences at their best are theories of pure and applied intentionality.

# SIX
## THE FREEDOM OF THE WILL

In these pages, I have tried to answer what to me are some of the most worrisome questions about how we as human beings fit into the rest of the universe. Our conception of ourselves as free agents is fundamental to our overall self-conception. Now, ideally, I would like to be able to keep both my commonsense conceptions and my scientific beliefs. In the case of the relation between mind and body, for example, I was able to do that. But when it comes to the question of freedom and determinism, I am – like a lot of other philosophers – unable to reconcile the two.

One would think that after over 2000 years of worrying about it, the problem of the freedom of the will would by now have been finally solved. Well, actually most philosophers think it has been solved. They think it was solved by Thomas Hobbes and David Hume and various other empirically-minded philosophers whose solutions have been repeated and improved right into the twentieth century. I think it has not been solved. In this lecture I want to give you an account of what the problem is, and why the contemporary solution is not a solution, and then conclude by trying to explain why the problem is likely to stay with us.

On the one hand we are inclined to say that since nature consists of particles and their relations with each other, and since everything can be accounted for in terms of those particles and their relations, there is simply no room for freedom of the will. As far as human freedom is concerned, it doesn't matter whether physics is deterministic, as Newtonian physics was, or whether it allows for an indeterminacy at the level of particle physics, as contemporary quantum mechanics does.

Indeterminism at the level of particles in physics is really no support at all to any doctrine of the freedom of the will; because first, the statistical indeterminacy at the level of particles does not show any indeterminacy at the level of the objects that matter to us – human bodies, for example. And secondly, even if there is an element of indeterminacy in the behaviour of physical particles – even if they are only statistically predictable – still, that by itself gives no scope for human freedom of the will; because it doesn't follow from the fact that particles are only statistically determined that the human mind can force the statistically-determined particles to swerve from their paths. Indeterminism is no evidence that there is or could be some mental energy of human freedom that can move molecules in directions that they were not otherwise going to move. So it really does look as if everything we know about physics forces us to some form of denial of human freedom.

The strongest image for conveying this conception of determinism is still that formulated by Laplace: If an ideal observer knew the positions of all the particles at a given instant and knew all the laws governing their movements, he could predict and retrodict the entire history of the universe. Some of the predictions of a contemporary quantum-mechanical Laplace might be statistical, but they would still allow no room for freedom of the will.

So much for the appeal of determinism. Now let's turn to the argument for the freedom of the will. As many philosophers have pointed out, if there is any fact of experience that we are all familiar with, it's the simple fact that our own choices, decisions, reasonings, and cogitations seem to make a difference to our actual behaviour. There are all sorts of experiences that we have in life where it seems just a fact of our experience that though we did one thing, we feel we know perfectly well that we could have done something else. We know we could have done something else, because we chose one thing for certain reasons. But we were aware that there were also reasons for choosing something else, and indeed, we might have acted on those reasons and chosen that something else. Another way to put this point is to say: it is just a plain empirical fact about our behaviour that it isn't predictable in the way that the behaviour of objects rolling down an inclined plane is predictable. And the reason it isn't predictable in that way is that we could often have done otherwise than we in fact did. Human freedom is just a fact of experience. If we want some empirical proof of this fact, we can simply point to the further fact that it is always up to us to falsify any predictions anybody might care to make about our behaviour. If somebody predicts that I am going to do something, I might just damn well do something else. Now, that sort of option is simply not open to glaciers moving down mountainsides or balls rolling down inclined planes or the planets moving in their elliptical orbits.

This is a characteristic philosophical conundrum. On the one hand, a set of very powerful arguments force us to the conclusion that free will has no place in the universe. On the other hand, a series of powerful arguments based on facts of our own experience inclines us to the conclusion that there must be some freedom of the will because we all experience it all the time.

There is a standard solution to this philosophical conundrum. According to this solution, free will and determinism are perfectly compatible with each other. Of course, everything in the world is determined, but some human actions are nonetheless free. To say that they are free is not to deny that they are determined; it is just to say that they are not constrained. We are not forced to do them, So, for example, if a man is forced to do something at gunpoint, or if he is suffering from some psychological compulsion, then his behaviour is genuinely unfree. But if on the other hand he freely acts, if he acts, as we say, of his own free will, then his behaviour is free. Of course it is also completely determined, since every aspect of his behaviour is determined by the physical forces operating on the particles that compose his body, as they operate on all of the bodies in the universe. So, free behaviour exists, but it is just a small corner of the determined world – it is that corner of determined human behaviour where certain kinds of force and compulsion are absent.

Now, because this view asserts the compatibility of free will and determinism, it is usually called simply 'compatibilism'. I think it is inadequate as a solution to the problem, and here is why. The problem about the freedom of the will is not about whether or not there are inner psychological reasons that cause us to do things as well as external physical causes and inner compulsions. Rather, it is about whether or not the causes of our behaviour, whatever they are, are sufficient to *determine* the behaviour so that things *have to* happen the way they do happen.

There's another way to put this problem. Is it ever true to say of a person that he could have done otherwise, all other conditions remaining the same? For example, given that a person chose to vote for the Tories, could he have chosen to vote for one of the other parties, all other conditions remaining the same? Now compatibilism doesn't really answer that question in a way that allows any scope for the ordinary notion of the freedom of the will. What it says is that all behaviour is determined in such a way that it couldn't have occurred otherwise, all other conditions remaining the same. Everything that happened was indeed determined. It's just that some things were determined by certain sorts of inner psychological causes (those which we call our 'reasons for acting') and not by external forces or psychological compulsions. So, we are still left with a problem. Is it ever true to say of a human being that he could have done otherwise?

The problem about compatibilism, then, is that it doesn't answer the question, 'Could we have done otherwise, all other conditions remaining the same?', in a way that is consistent with our belief in our own free will. Compatibilism, in short, denies the substance of free will while maintaining its verbal shell.

Let us try, then, to make a fresh start. I said that we have a conviction of our own free will simply based on the facts of human experience. But how reliable are those experiences? As I mentioned earlier, the typical case, often described by philosophers, which inclines us to believe in our own free will is a case where we confront a bunch of choices, reason about what is the best thing to do, make up our minds, and then do the thing we have decided to do.

But perhaps our belief that such experiences support the doctrine of human freedom is illusory. Consider this sort of example. A typical hypnosis experiment has the following form. Under hypnosis the patient is given a post-hypnotic suggestion. You can tell him, for example, to do some fairly trivial, harmless thing, such as, let's say, crawl around on the floor. After the patient comes out of hypnosis, he might be engaging in conversation, sitting, drinking coffee, when suddenly he says something like, 'What a fascinating floor in this room!', or 'I want to check out this rug', or 'I'm thinking of investing in floor coverings and I'd like to investigate this floor.' He then proceeds to crawl around on the floor. Now the interest of these cases is that the patient always gives some more or less adequate reason for doing what he does. That is, he seems to himself to be behaving freely. We, on the other hand, have good reasons to believe that his behaviour isn't free at all, that the reasons he gives for his apparent decision to crawl around on the floor are irrelevant, that his behaviour was determined in advance, that in fact he is in the grip of a post-hypnotic suggestion. Anybody who knew the facts about him could have predicted his behaviour in advance. Now, one way to pose the problem of determinism, or at least one aspect of the problem of determinism, is: 'Is all human behaviour like that?' Is all human behaviour like the man operating under a post-hypnotic suggestion?

But now if we take the example seriously, it looks as if it proves to be an argument for the freedom of the will and not against it. The agent thought he was acting freely, though in fact his behaviour was determined. But it seems empirically very unlikely that all human behaviour is like that. Sometimes people are suffering from the effects of hypnosis, and sometimes we know that they are in the grip of unconscious urges which they cannot control. But are they always like that? Is all behaviour determined by such *psychological* compulsions? If we try to treat psychological determinism as a factual claim about our behaviour, then it seems to be just plain false. The thesis of psychological determinism is that prior psychological causes determine all of our behaviour in the way that they determine the behaviour of the hypnosis subject or the heroin addict. On this view, all behaviour, in one way or another, is psychologically compulsive. But the available evidence suggests that such a thesis is false. We do indeed normally act on the basis of our intentional states – our beliefs, hopes, fears, desires, etc. – and in that sense our mental states function causally. But this form of cause and effect is not deterministic. We might have had exactly those mental states and still not have done what we did. As far as psychological causes are concerned, we could have done otherwise. Instances of hypnosis and psychologically compulsive behaviour on the other hand are usually pathological and easily distinguishable from normal free action. So, psychologically speaking, there is scope for human freedom.

But is this solution really an advance on compatibilism? Aren't we just saying, once again, that yes, all behaviour is determined, but what we call free behaviour is the sort determined by rational thought processes? Sometimes the conscious, rational thought processes don't make any difference, as in the hypnosis case, and sometimes they do, as in the normal case. Normal cases are those where we say the agent is really free. But of course those normal rational thought processes are as much determined as anything else. So once again, don't we have the result that everything we do was entirely written in the book of history billions of years before we were born, and therefore, nothing we do is free in any philosophically interesting sense? If we choose to call our behaviour free, that is just a matter of adopting a traditional terminology. Just as we continue to speak of 'sunsets' even though we know the sun doesn't literally set; so we continue to speak of 'acting of our own free will' even though there is no such phenomenon.

One way to examine a philosophical thesis, or any other kind of a thesis for that matter, is to ask, 'What difference would it make? How would the world be any different if that thesis were true as opposed to how the world would be if that thesis were false?' Part of the appeal of determinism, I believe, is that it seems to be consistent with the way the world in fact proceeds, at least as far as we know anything about it from physics. That is, if determinism were true, then the world would proceed pretty much the way it does proceed, the only difference being that certain of our beliefs about its proceedings would be false. Those beliefs are important to us because they have to do with the belief that we could have done things differently from the way we did in fact do them. And this belief in turn connects with beliefs about moral responsibility and our own nature as persons. But if libertarianism, which is the thesis of free will, were true, it appears we would have to make some really radical changes in our beliefs about the world. In order for us to have radical freedom, it looks as if we would have to postulate that inside each of us was a self that was capable of interfering with the causal order of nature. That is, it looks as if we would have to contain some entity that was capable of making molecules swerve from their paths. I don't know if such a view is even intelligible, but it's certainly not consistent with what we know about how the world works from physics. And there is not the slightest evidence to suppose that we should abandon physical theory in favour of such a view.

So far, then, we seem to be getting exactly nowhere in our effort to resolve the conflict between determinism and the belief in the freedom of the will. Science allows no place for the freedom of the will, and indeterminism in physics offers no support for it. On the other hand, we are unable to give up the belief in the freedom of the will. Let us investigate both of these points a bit further.

Why exactly is there no room for the freedom of the will on the contemporary scientific view? Our basic explanatory mechanisms in physics work from the bottom up. That is to say, we explain the behaviour of surface features of a phenomenon such as the transparency of glass or the liquidity of water, in terms of the behaviour of microparticles such as molecules. And the relation of the mind to the brain is an example of such a relation. Mental features are caused by, and realised in neurophysiological phenomena, as I discussed in the first chapter. But we get causation from the mind to the body, that is we get top-down causation over a passage of time; and we get top-down causation over time because the top level and the bottom level go together. So, for example, suppose I wish to cause the release of the neurotransmitter acetylcholine at the axon end-plates of my motorneurons, I can do it by simply deciding to raise my arm and then raising it. Here, the mental event, the intention to raise my arm, causes the physical event, the release of acetylcholine – a case of top-down causation if ever there was one. But the top-down causation works only because the mental events are grounded in the neurophysiology to start with. So, corresponding to the description of the causal relations that go from the top to the bottom, there is another description of the same series of events where the causal relations bounce entirely along the bottom, that is, they are entirely a matter of neurons and neuron firings at synapses, etc. As long as we accept this conception of how nature works, then it doesn't seem that there is any scope for the freedom of the will because on this conception the mind can only affect nature in so far as it is a part of nature. But if so, then like the rest of nature, its features are determined at the basic micro-levels of physics.

This is an absolutely fundamental point in this chapter, so let me repeat it. The form of determinism that is ultimately worrisome is not psychological determinism. The idea that our states of mind are sufficient to determine everything we do is probably just false. The worrisome form of determinism is more basic and fundamental. Since all of the surface features of the world are entirely caused by and realised in systems of micro-elements, the behaviour of micro-elements is sufficient to determine everything that happens. Such a 'bottom up' picture of the world allows for top-down causation (our minds, for example, can affect our bodies). But top-down causation only works because the top level is already caused by and realised in the bottom levels.

Well then, let's turn to the next obvious question. What is it about our experience that makes it impossible for us to abandon the belief in the freedom of the will? If freedom is an illusion, why is it an illusion we seem unable to abandon? The first thing to notice about our conception of human freedom is that it is essentially tied to consciousness. We only attribute freedom to conscious beings. If, for example, somebody built a robot which we believed to be totally unconscious, we would never feel any inclination to call it free. Even if we found its behaviour random and unpredictable, we would not say that it was acting freely in the sense that we think of ourselves as acting freely. If on the other hand somebody built a robot that we became convinced had consciousness, in the same sense that we do, then it would at least be an open question whether or not that robot had freedom of the will.

The second point to note is that it is not just any state of the consciousness that gives us the conviction of human freedom. If life consisted entirely of the reception of passive perceptions, then it seems to me we would never so much as form the idea of human freedom. If you imagine yourself totally immobile, totally unable to move, and unable even to determine the course of your own thoughts, but still receiving stimuli, for example, periodic mildly painful sensations, there would not be the slightest inclination to conclude that you have freedom of the will.

I said earlier that most philosophers think that the conviction of human freedom is somehow essentially tied to the process of rational decision-making. But I think that is only partially true. In fact, weighing up reasons is only a very special case of the experience that gives us the conviction of freedom. The characteristic experience that gives us the conviction of human freedom, and it is an experience from which we are unable to strip away the conviction of freedom, is the experience of engaging in voluntary, intentional human actions. In our discussion of intentionality we concentrated on that form of intentionality which consisted in conscious intentions in action, intentionality which is causal in the way that I described, and whose conditions of satisfaction are that certain bodily movements occur, and that they occur as caused by that very intention in action. It is this experience which is the foundation stone of our belief in the freedom of the will. Why? Reflect very carefully on the character of the experiences you have as you engage in normal, everyday ordinary human actions. You will sense the possibility of alternative courses of action built into these experiences. Raise your arm or walk across the room or take a drink of water, and you will see that at any point in the experience you have a sense of alternative courses of action open to you.

If one tried to express it in words, the difference between the experience of perceiving and the experience of acting is that in perceiving one has the sense: 'This is happening to me,' and in acting one has the sense: 'I am making this happen.' But the sense that 'I am making this happen' carries with it the sense that 'I could be doing something else'. In normal behaviour, each thing we do carries the conviction, valid or invalid, that we could be doing something else right here and now, that is, all other conditions remaining the same. This, I submit, is the source of our unshakable conviction of our own free will. It is perhaps important to emphasise that I am discussing normal human action. If one is in the grip of a great passion, if one is in a great rage, for example, one loses this sense of freedom and one can even be surprised to discover what one is doing.

Once we notice this feature of the experience of acting, a great many of the puzzling phenomena I mentioned earlier are easily explained. Why for example do we feel that the man in the case of post-hypnotic suggestion is not acting freely in the sense in which we are, even though he might think that he is acting freely? The reason is that in an important sense he doesn't know what he is doing. His actual intention-in-action is totally unconscious. The options that he sees as available to him are irrelevant to the actual motivation of his action. Notice also that the compatibilist examples of 'forced' behaviour still, in many cases, involve the experience of freedom. If somebody tells me to do something at gunpoint, even in such a case I have an experience which has the sense of alternative courses of action built into it. If, for example, I am instructed to walk across the room at gunpoint, still part of the experience is that I sense that it is literally open to me at any step to do something else. The experience of freedom is thus an essential component of any case of acting with an intention.

Again, you can see this if you contrast the normal case of action with the Penfield cases, where stimulation of the motor cortex produces an involuntary movement of the arm or leg. In such a case the patient experiences the movement passively, as he would experience a sound or a sensation of pain. Unlike intentional actions, there are no options built into the experience. To see this point clearly, try to imagine that a portion of your life was like the Penfield experiments on a grand scale. Instead of walking across the room you simply find that your body is moving across the room; instead of speaking you simply hear and feel words coming out of your mouth. Imagine your experiences are those of a purely passive but conscious puppet and you will have imagined away the experience of freedom. But in the typical case of intentional action, there is no way we can carve off the experience of freedom. It is an essential part of the experience of acting.

This also explains, I believe, why we cannot give up our conviction of freedom. We find it easy to give up the conviction that the earth is flat as soon as we understand the evidence for the heliocentric theory of the solar system. Similarly when we look at a sunset, in spite of appearances we do not feel compelled to believe that the sun is setting behind the earth, we believe that the appearance of the sun setting is simply an illusion created by the rotation of the earth. In each case it is possible to give up a commonsense conviction because the hypothesis that replaces it both accounts for the experiences that led to that conviction in the first place as well as explaining a whole lot of other facts that the commonsense view is unable to account for. That is why we gave up the belief in a flat earth and literal 'sunsets' in favour of the Copernican conception of the solar system. But we can't similarly give up the conviction of freedom because that conviction is built into every normal, conscious intentional action. And we use this conviction in identifying and explaining actions. This sense of freedom is not just a feature of deliberation, but is part of any action, whether premeditated or spontaneous. The point has nothing essentially to do with deliberation; deliberation is simply a special case.

We don't navigate the earth on the assumption of a flat earth, even though the earth looks flat, but we do act on the assumption of freedom. In fact we can't act otherwise than on the assumption of freedom, no matter how much we learn about how the world works as a determined physical system.

We can now draw the conclusions that are implicit in this discussion. First, if the worry about determinism is a worry that all of our behaviour is in fact psychologically compulsive, then it appears that the worry is unwarranted. Insofar as psychological determinism is an empirical hypothesis like any other, then the evidence we presently have available to us suggests it is false. Thus, this does give us a modified form of compatibilism. It gives us the view that psychological libertarianism is compatible with physical determinism.

Secondly, it even gives us a sense of 'could have' in which people's behaviour, though determined, is such that in that sense they could have done otherwise: The sense is simply that as far as the *psychological* factors were concerned, they could have done otherwise. The notions of ability, of what we are able to do and what we could have done, are often relative to some such set of criteria. For example, I could have voted for Carter in the 1980 American election, even if I did not; but I could not have voted for George Washington. He was not a candidate. So there is a sense of 'could have', in which there were a range of choices available to me, and in that sense there were a lot of things I could have done, all other things being equal, which I did not do. Similarly, because the psychological factors operating on me do not always, or even in general, compel me to behave in a particular fashion, I often, psychologically speaking, could have done something different from what I did in fact do.

But third, this form of compatibilism still does not give us anything like the resolution of the conflict between freedom and determinism that our urge to radical libertarianism really demands. As long as we accept the bottom-up conception of physical explanation, and it is a conception on which the past three hundred years of science are based, then psychological facts about ourselves, like any other higher level facts, are entirely causally explicable in terms of and entirely realised in systems of elements at the fundamental micro-physical level. Our conception of physical reality simply does not allow for radical freedom.

Fourth, and finally, for reasons I don't really understand, evolution has given us a form of experience of voluntary action where the experience of freedom, that is to say, the experience of the sense of alternative possibilities, is built into the very structure of conscious, voluntary, intentional human behaviour. For that reason, I believe, neither this discussion nor any other will ever convince us that our behaviour is unfree.

My aim in this book has been to try to characterise the relationships between the conception that we have of ourselves as rational, free, conscious, mindful agents with a conception that we have of the world as consisting of mindless, meaningless, physical particles. It is tempting to think that just as we have discovered that large portions of common sense do not adequately represent how the world really works, so we might discover that our conception of ourselves and our behaviour is entirely false. But there are limits on this possibility. The distinction between reality and appearance cannot apply to the very existence of consciousness. For if it seems to me that I'm conscious, I *am* conscious. We could discover all kinds of startling things about ourselves and our behaviour; but we cannot discover that we do not have minds, that they do not contain conscious, subjective, intentionalistic mental states; nor could we discover that we do not at least try to engage in voluntary, free, intentional actions. The problem I have set myself is not to prove the existence of these things, but to examine their status and their implications for our conceptions of the rest of nature. My general theme has been that, with certain important exceptions, our commonsense mentalistic conception of ourselves is perfectly consistent with our conception of nature as a physical system.

# SUGGESTIONS FOR FURTHER READING

BLOCK, NED (ed.), *Readings in Philosophy and Psychology*, Vols. 1 & 2, Cambridge: Harvard University Press, 1981.

DAVIDSON, DONALD, *Essays on Actions and Events*, Oxford: Oxford University Press, 1980

DREYFUS, HUBERT L., *What Computers Can't Do: The Limits of Artificial Intelligence*, New York: Harper & Row, 1979 (revised).

FODOR, JERRY, *Representations: Philosophical Essays on the Foundations of Cognitive Science*, Cambridge: MIT Press, 1983.

HAUGELAND, JOHN (ed.), *Mind Design*, Cambridge: MIT Press, 1981.

KUFFLER, STEPHEN W. & NICHOLAS, JOHN G., *From Neuron to Brain: A Cellular Approach to the Function of the Nervous System*, Sunderland, Mass.: Sinauer Associates, 1976.

MORGENBESSER, SYDNEY & WALSH, JAMES (eds.), *Free Will*, Englewood Cliffs: Prentice-Hall, Inc., 1962.

NAGEL, THOMAS, *Mortal Questions*, Cambridge: Cambridge University Press, 1979.

NORMAN, DONALD A. (ed.), *Perspectives on Cognitive Science*, Norwood: Albex Publishing Corp., 1981.

PENFIELD, WILDER, *The Mystery of the Mind*, Princeton: Princeton University Press, 1975.

ROSENZWEIG, MARK & LEIMAN, ARNOLD, *Physiological Psychology*, Lexington, Mass.: D. C. Heath & Co., 1982.

SEARLE, JOHN R., *Intentionality: An Essay in the Philosophy of Mind*, Cambridge: Cambridge University Press, 1983.

SHEPHERD, GORDON M., *Neurobiology*, Oxford: Oxford University Press, 1983.

WHITE, ALAN R. (ed.), *The Philosophy of Action*, Oxford: Oxford University Press, 1968.

# INDEX

BOOK
CARL SANDBURG COLLEGE
science Searle, John